Vue.js 3
By Example

Blueprints to learn Vue web development,
full-stack development, and cross-platform
development quickly

John Au-Yeung

BIRMINGHAM—MUMBAI

Vue.js 3 By Example

Copyright © 2021 Packt Publishing

Publishing Product Manager: Kaustubh Manglurkar
Senior Editor: Keagan Carneiro
Content Development Editor: Abhishek Jadhav
Technical Editor: Deepesh Patel
Copy Editor: Safis Editing
Project Coordinator: Manthan Patel
Proofreader: Safis Editing
Indexer: Vinayak Purushotham
Production Designer: Roshan Kawale

First published: April 2021
Production reference: 1230421

Published by Packt Publishing Ltd.
Livery Place
35 Livery Street
Birmingham
B3 2PB, UK.

ISBN 978-1-83882-634-5

www.packt.com

Contributors

About the author

John Au-Yeung is a frontend developer who has extensive experience with the latest frontend technologies.

He has an MSc in information technology and a BSc in mathematics.

He is also a part-time blogger who writes about the latest frontend development technologies. In addition, he is an author of many self-published books about JavaScript programming. He has extensive experience with Vue and React and loves working with both. JavaScript programming is his focus and is what he does every day.

About the reviewer

Jaime Jones is a software engineer with a passion for frontend development, mostly focused on Vue.js. She graduated from Boise State University with a degree in psychology, but then decided to turn her hobby into her career and pursued development instead. She works as a lead frontend developer at Ritter Insurance Marketing and enjoys speaking at conferences and writing blog posts to contribute to the community.

Table of Contents

4
Building a Photo Management Desktop App

5
Building a Multipurpose Calculator Mobile App with Ionic

6

Building a Vacation Booking App with the PrimeVue UI Framework

7

Creating a Shopping Cart System with GraphQL

8

Building a Chat App with Vue 3, Laravel, and Socket.IO

Other Books You May Enjoy

Index

Preface

Vue is one of the leading frameworks with a huge ecosystem and increasing adoption due to its ease of use when developing applications and the fact that it can help you achieve impressive results during development quickly. This book explores the latest Vue version – Vue 3.0 – and how you can leverage it effectively.

You'll learn with the help of an example-based approach, starting with exploring the basics of Vue 3 by creating a simple application and looking at features such as components, directives, and their usage. To build your knowledge and enable you to have confidence in your app-building skills, the book will show you how to test that app with Jest and Vue Test Utils. Later, you'll learn how to write non-web apps with Vue 3 and create cross-platform desktop apps with the Electron plugin. You'll also learn how to create a multi-purpose mobile app with Vue and Ionic. As you progress, you'll learn how to develop web apps with Vue 3 that interact well with GraphQL APIs. Finally, you'll build a real-time chat app that performs real-time communication using Vue 3 and Laravel.

By the end of this book, you'll have developed the real-world skills you need by working through a range of app-building projects using Vue 3.

Who this book is for

This book is for web developers who are interested in learning frontend web development with Vue 3 and creating professional applications using it. You will also find this book useful if you want to learn how to create full-stack web apps with Vue.js 3.0 as the frontend. Prior knowledge of basic JavaScript programming is required to get the most out of this book.

What this book covers

Chapter 1, *Creating Your First Application in Vue 3*, will look at how to use Vue 3 to create simple apps. You will start by building the most basic apps and then move on to building more complex solutions.

Chapter 2, Building a Vue 3 Progressive Web App, will teach you how to create a GitHub **Progressive Web App** (**PWA**) with Vue 3. As you build the project, you will examine the inner workings of Vue apps by looking at the basic building blocks in depth. You will create Vue apps with components and will look at the parts that make up a component and how they work.

Chapter 3, Building a Slider Puzzle Game with Tests, will introduce you to Vue by having you create a simple game with Vue 3. You will learn how to use different methods, mixins, directives, and computed properties to be added to the project.

Chapter 4, Building a Photo Management Desktop App, will help you build a photo management desktop app with the Vue Electron plugin. You will learn how to build cross-platform desktop apps easily with Electron and Vue.

Chapter 5, Building a Multipurpose Calculator Mobile App with Ionic, will see you create a multi-purpose calculator mobile app with NativeScript. You will use Vuex to manage the state and save results data so that you can use it later in local storage. Finally, you will master currency conversions, unit conversions, and tips calculations.

Chapter 6, Building a Vacation Booking App with the PrimeVue UI Framework, will teach you how to create a travel booking app with admin functionality. The admin side will be a dashboard for users to manage bookings. It will involve using state management and routing to create a full-fledged app. The backend will be simple so that you can focus more on Vue. State management with Vuex and routing with Vue Router will also be required.

Chapter 7, Creating a Shopping Cart System with GraphQL, will help you create a Vue 3 app and use it with GraphQL APIs. You will learn how to use a GraphQL client within our Vue 3 app. The API will have queries, mutations, and database interactions, and you will learn how to create a GraphQL API with Express.

Chapter 8, Building a Chat App with Vue 3, Laravel, and Socket.IO, will teach you how to create a chat app with Vue 3, socket.io, and Laravel. This app will make HTTP requests and have real-time communication. It can be used by multiple users.

To get the most out of this book

To get the most out of this book, you should know the basics of modern JavaScript. Knowing how to program with JavaScript features introduced from 2015 onward would make understanding this book much easier. Basic TypeScript concepts, such as defining interfaces and advanced data types, will be used in *Chapter 5, Building a Multipurpose Calculator Mobile App with Ionic*.

Also, *Chapter 8*, *Building a Chat App with Vue 3, Laravel, and Socket.IO*, covers Laravel, which requires a basic understanding of PHP. The backend portions of the more advanced projects also require understanding very basic SQL syntax. Commands such as `Select`, `Insert`, and `Create table` will be helpful.

Software/hardware covered in the book	OS requirements
Vue.js 3	Windows, macOS X, or Linux
TypeScript	
Vue CLI	
ECMAScript 6 and above	

The other things required are the latest versions of Node.js and Visual Studio Code. Visual Studio Code supports JavaScript and TypeScript development out of the box. Node.js is required to run Vue CLI and Ionic CLI.

After reading this book, you should try to practice more by creating your own projects. This way, you will utilize the knowledge that you gained from this book. Learning from tutorials is only a start; creating projects on your own will make you proficient.

If you are using the digital version of this book, we advise you to type the code yourself or access the code via the GitHub repository (link available in the next section). Doing so will help you avoid any potential errors related to the copying and pasting of code.

Download the example code files

You can download the example code files for this book from your account at www. packt.com. If you purchased this book elsewhere, you can visit www.packtpub.com/ support and register to have the files emailed directly to you.

You can download the code files by following these steps:

1. Log in or register at www.packt.com.
2. Select the **Support** tab.
3. Click on **Code Downloads**.
4. Enter the name of the book in the **Search** box and follow the onscreen instructions.

Once the file is downloaded, please make sure that you unzip or extract the folder using the latest version of:

- WinRAR/7-Zip for Windows
- Zipeg/iZip/UnRarX for Mac
- 7-Zip/PeaZip for Linux

The code bundle for the book is also hosted on GitHub at `https://github.com/PacktPublishing/-Vue.js-3-By-Example`. In case there's an update to the code, it will be updated on the existing GitHub repository.

We also have other code bundles from our rich catalog of books and videos available at `https://github.com/PacktPublishing/`. Check them out!

Conventions used

There are a number of text conventions used throughout this book.

`Code in text`: Indicates code words in text, database table names, folder names, filenames, file extensions, pathnames, dummy URLs, user input, and Twitter handles. Here is an example: "Mount the downloaded `WebStorm-10*.dmg` disk image file as another disk in your system."

A block of code is set as follows:

```
html, body, #map {
  height: 100%;
  margin: 0;
  padding: 0
}
```

When we wish to draw your attention to a particular part of a code block, the relevant lines or items are set in bold:

```
[default]
exten => s,1,Dial(Zap/1|30)
exten => s,2,Voicemail(u100)
exten => s,102,Voicemail(b100)
exten => i,1,Voicemail(s0)
```

Any command-line input or output is written as follows:

```
$ mkdir css
$ cd css
```

Bold: Indicates a new term, an important word, or words that you see onscreen. For example, words in menus or dialog boxes appear in the text like this. Here is an example: "Select **System info** from the **Administration** panel."

Tips or important notes
Appear like this.

Get in touch

Feedback from our readers is always welcome.

General feedback: If you have questions about any aspect of this book, mention the book title in the subject of your message and email us at customercare@packtpub.com.

Errata: Although we have taken every care to ensure the accuracy of our content, mistakes do happen. If you have found a mistake in this book, we would be grateful if you would report this to us. Please visit www.packtpub.com/support/errata, selecting your book, clicking on the Errata Submission Form link, and entering the details.

Piracy: If you come across any illegal copies of our works in any form on the Internet, we would be grateful if you would provide us with the location address or website name. Please contact us at copyright@packt.com with a link to the material.

If you are interested in becoming an author: If there is a topic that you have expertise in and you are interested in either writing or contributing to a book, please visit authors.packtpub.com.

Reviews

Please leave a review. Once you have read and used this book, why not leave a review on the site that you purchased it from? Potential readers can then see and use your unbiased opinion to make purchase decisions, we at Packt can understand what you think about our products, and our authors can see your feedback on their book. Thank you!

For more information about Packt, please visit packt.com.

1
Creating Your First Application in Vue 3

Vue 3 is the latest version of the popular Vue.js framework. It is focused on improving developer experience and speed. It is a component-based framework that lets us create modular, testable apps with ease. It includes concepts that are common to other frameworks such as props, transitions, event handling, templates, directives, data binding, and more. The main goal of this chapter is to get you started with developing your first Vue app. This chapter is focused on how to create components.

In this chapter, we will look at how to use Vue 3 to create simple apps from scratch. We will start by building the most basic apps and then move on to building more complex solutions in the next few chapters.

The major topics we will cover are as follows:

- Understanding Vue as a framework
- Setting up the Vue project
- Vue 3 core features – components and built-in directives
- Debugging with Vue.js Devtools

Technical requirements

The code for this chapter is located at `https://github.com/PacktPublishing/-Vue.js-3-By-Example/tree/master/Chapter01`.

Understanding Vue as a framework

As we mentioned in the introduction, there are concepts in Vue that are available from other frameworks. Directives manipulate the **Document Object Model (DOM)** just like in Angular.js and Angular. Templates render data like we do with Angular. It also has its own special syntax for data binding and adding directives.

Angular and React both have props that pass data between components. We can also loop through array and object entries to display items from lists. Also, like Angular, we can add plugins to a Vue project to extend its functionality.

Concepts that are exclusive to Vue.js include computed properties, which are component properties that are derived from other properties. Also, Vue components have watchers that let us watch for reactive data changes. Reactive data is data that is watched by Vue.js and actions are done automatically when reactive data changes.

As reactive data changes, other parts of a component and other components that reference those values are all updated automatically. This is the magic of Vue. It is one of the reasons that we can do so much with so little code. It takes care of the task of watching for data changes for us, so that we don't have to do that.

Another unique feature of Vue 3 is that we can add the framework and its libraries with script tags. This means that if we have a legacy frontend, we can still use Vue 3 and its libraries to enhance legacy frontends. Also, we don't need to add build tools to build our app. This is a great feature that isn't available with most other popular frameworks.

There's also the popular Vue Router library for routing, and the Vuex library for state management. They have all been updated to be compatible with Vue 3, so we can use them safely. This way, we don't have to worry about which router and state management library to use as we do with other frameworks such as React. Angular comes with its own routes, but no standard state management library has been designated.

Setting up the Vue project with the Vue CLI and script tag

There are several ways to create Vue projects or to add script tags to our existing frontends. For prototyping or learning purposes, we can add the latest version of Vue 3 by adding the following code:

```
<script src="https://unpkg.com/vue@next"></script>
```

This will always include the latest version of Vue in our app. If we use it in production, we should include the version number to avoid unexpected changes from newer versions breaking our app. The version number can replace the next word if we want to specify the version.

We can also install Vue by installing it as a package. To do that, we can run the following command:

```
npm install vue@next
```

This will install the latest version of Vue in our JavaScript project.

If we created a Vue project from scratch with an older version of the Vue CLI, then we can use the CLI to generate all the files and install all the packages for us. This is the easiest way to get started with a Vue project. With Vue 3, we should use Vue CLI v4.5 by running the following command:

```
yarn global add @vue/cli@next
```

We can also install the Vue palate by running the following command:

```
npm install -g @vue/cli@next
```

Then, to upgrade our Vue project to Vue 3, we can run the following command:

```
vue upgrade --next
```

The Vite build tool will let us create a Vue 3 project from scratch. It lets us serve our project much faster than with the Vue CLI because it can work with modules natively. We can set up a Vue project from scratch by running these commands with NPM:

```
$ npm init vite-app <project-name>
$ cd <project-name>
$ npm install
$ npm run dev
```

With Yarn, we must run the following commands:

```
$ yarn create vite-app <project-name>
$ cd <project-name>
$ yarn
$ yarn dev
```

In either case, we replace `<project-name>` with the project name of our choice.

There are various builds of Vue that we can use. One set are CDN versions, which don't come with bundlers. We can recognize them by the `vue(.runtime).global(.prod).js` pattern in the filename. These can be included directly with script tags.

We use them with templates that are directly added to the HTML. The `vue.global.js` file is the full build and includes both the compiler and the runtime, so it can compile templates on the fly from HTML. The `vue.runtime.global.js` file only contains the runtime and requires the template to be precompiled during a build step.

The development and production branches are hardcoded, and we can tell them apart by checking if the file ends with `.prod.js`. These files are production ready as they're minified. These aren't **Universal Module Definition (UMD)** builds. They contain IIFEs that are meant to be used with regular script tags.

If we use a bundler such as Webpack, Rollup, or Parcel, then we can use the `vue(.runtime).esm-bundler.js` file. The development and production branches are determined by the `process.env.NODE_ENV` property. It also has the full version, which compiles the template on the fly at runtime and a runtime version.

In this chapter, we will be going through the basic features of Vue with the script tag version of Vue. In the subsequent chapters, we'll move on to using the Vue CLI to create our Vue 3 projects. This way, we can focus on exploring the basic features of Vue 3, which will be handy when we move on to creating more complex projects. Let's begin by creating a Vue instance.

Creating your Vue instance

Now that we have set up our Vue project, we can look at the Vue instance more closely. All Vue 3 apps have a Vue instance. The Vue instance serves as the entry point of the app. This means this is what is loaded first. It is the root component of the app, and it has a template and a component option object to control how the template is rendered in the browser.

To create our first Vue 3 app, we must add the following code to the `index.html` file:

```html
<!DOCTYPE html>
<html lang="en">
  <head>
    <title>Vue App</title>
    <script src="https://unpkg.com/vue@next"></script>
  </head>
  <body>
    <div id="app">
      count: {{ count }}
    </div>
    <script>
      const Counter = {
        data() {
          return {
            count: 0
          };
        }
      };

      Vue.createApp(Counter).mount("#app");
    </script>
  </body>
</html>
```

In our first Vue 3 app, we started by adding the `script` tag to add the Vue framework script. It's not final yet, so we added the next version of the Vue script.

In the body, we have a `div` with the ID app, which we use to hold the template. The only content that is inside the template will be compiled by the template compiler that comes with Vue 3. Below that, we have a `script` tag to create our app. It provides the `Counter` object, which contains the properties we can use to create our app.

Vue components come as objects that will be used by Vue to create any necessary components. The `data` property is a special property that returns the initial values of our states. The states are automatically reactive. The `count` state is a reactive state that we can update. It is the same one that's in the template. Anything in the curly braces must be some expression that contains reactive properties or other JavaScript expressions.

If we add reactive states between the curly braces, then they will be updated. Since the `count` reactive property is initialized to `0`, the `count` property is also `0` in the template. The `Counter` object is passed into the `Vue.createApp` method to compile the template and connect the reactive properties, to render the expression inside the curly braces as the final result. So, we should see `count:   0` in our rendered output.

The `mount()` method accepts a CSS selector string as its argument. The selector is the template to render the app in. Whatever is inside it will be considered Vue expressions, and they will be rendered accordingly. Expressions in curly braces will be rendered and attributes will be interpreted by Vue as props or directives, depending on how they are written.

In the next section, we will look at the core features of Vue.js 3.

Vue 3 core features – components and built-in directives

Now that we have created a basic Vue app with a Vue instance, we can look more closely at how to make it do more. Vue 3 is a component-based framework. Therefore, components are the core building blocks that are used to build full production - quality Vue 3 apps. Components are parts that can be combined to form a full app and are reusable. Vue 3 components have several parts, which include the template, the component option object, and the styles. The styles are the CSS styles that we apply to the rendered elements. The template is what is rendered on the browser's screen. It contains HTML code combined with JavaScript expressions to form the content that's rendered in the browser.

Templates get their data from the corresponding component option object. Also, the component templates can have directives that control how content is rendered and how to bind data from the template to a reactive property.

Components

We created a basic Vue app with a Vue instance. Now, we must find a way to organize our app in a manageable way. Vue 3 is a component-based frontend framework. This means that apps created with Vue 3 are created by composing multiple components into one. This way, we can keep each part of our app small, and this helps with making testing easy, as well as easy to debug. This is something that is important as we are creating a non-trivial app that provides functionality for users.

In Vue 3, a component is a Vue instance with some predefined options. To use components in another component, we must register them. To create a Vue component, we can call the `app.component()` method. The first argument is the component, called `string`, while the second argument is an object that contains the component options.

A minimal component should at least contain the template property that was added to the object. This way, it will display something in our component to make it useful. We will start by creating a component for displaying todo items. To display our todo item, we can create a `todo-item` component. Also, a component most likely needs to accept props to display data from its parent component. A **prop** is a special attribute that lets a Vue component pass some data to a child component. A child component has the `props` property to define what kind of value it will accept. To do this, we can write the following code:

```html
<!DOCTYPE html>
<html lang="en">
  <head>
    <title>Vue App</title>
    <script src="https://unpkg.com/vue@next"></script>
  </head>
  <body>
    <div id="app">
      <div>
        <ol>
            ...
          ]
        };
      }
    };

    const app = Vue.createApp(App);

    app.component("todo-item", {
      props: ["todo"],
      template: `<li>{{todo.description}}</li>`
    });

    app.mount("#app");
```

```
        </script>
    </body>
</html>
```

We called the `app.component()` method to create the `todo-item` component. It contains the `props` property with an array of prop names to accept the `todo` prop. The way we defined the prop means that we can accept any value as the value of the `todo` prop. We can also specify them with a value type, set whether it is required or not, or provide a default value for it. The `template` property lets us render it when we want to. We just set it to a string, and it will render the items like any other template.

The `li` element is rendered in the template. The curly braces work the same way as any other template. It is used to interpolate the value. To access the prop's value, we just access it as a property of this in the component or just with the prop name itself in the template.

To pass the `todo` prop from the root Vue instance to the `todo-item` component, we prefix the prop name with a colon to indicate that it is a prop. The colon is short for `v-bind`. The `v-bind` directive is a built-in Vue directive that lets us pass data to a child component as a prop. If we have a prop name that is in camel case, then it will be mapped to a kebab-case name in the HTML to keep it valid. This is because valid attributes should have kebab-case names. The template compiler that comes with Vue 3 will do the mapping automatically. So, we just have to follow the conventions and then we can pass our props correctly.

If we are using the `v-for` directive, we should add the key prop so that Vue 3 can keep track of the items properly. With the `v-for` directive, we can loop through an array or object and display the entries from them. The value should be a unique ID so that Vue can render the items properly, even if we swap the positions of the items and add or delete items and perform other actions in a list. To do this, we can write the following code:

```
<!DOCTYPE html>
<html lang="en">
  <head>
    <title>Vue App</title>
    <script src="https://unpkg.com/vue@next"></script>
  </head>
  <body>
    ...
    </div>
    <script>
      const App = {
```

```
      data() {
        return {
          todos: [
            { id: 1, description: "eat" },
            { id: 2, description: "drink" },
            { id: 3, description: "sleep" }
          ]
        };
        ...


    app.mount("#app");
  </script>
 </body>
</html>
```

Each `id` property value is unique for Vue's list tracking to work.

Vue components look like custom elements in the web component's specification, but Vue components are not custom elements. They can't be used interchangeably. It is just a way to use a familiar syntax for creating components, and this is standard. There are some features in Vue components that are not available in custom elements. There is no cross-component data flow, custom event communication, and build tool integration with custom components. However, all these features are available in Vue components. We will cover these features of Vue components in the following sections.

Component life cycle

Each Vue 3 component has its own life cycle, and each life cycle stage has its own method. If the given stage of the life cycle is reached and if the method is defined in the component, the method will be run.

Right after the app is mounted with `app.mount()`, the events and life cycle are initialized. The first method that will be run when the component is being loaded is the `beforeCreate()` method. Then, the components are initialized with the reactive properties. Then, the `created()` method is run. Since the reactive properties are initialized at this stage, we can access the reactive properties in this method and the methods that are loaded after this one.

Then, the component's template or render functions are run to render the items. Once the content is loaded, `beforeMount` is run. Once `beforeMount` is run, the app will be mounted into the element that we specified with the selector we passed into the `app.mount()` method.

Once the app is mounted into the element, the mounted hook is run. Now, when any reactive property changes, the `beforeUpdate` hook is run. Then, the virtual DOM is rerendered, and the latest items are rendered from the latest values of the reactive properties. It is a good place to run any initialization code for outside libraries. Once that is done, the `updated` hook is run.

`beforeDestroy` is run right before the component is unmounted. It is a good place to run any cleanup code before destroying the component. The `destroyed` hook is run when the component is destroyed. The reactive properties won't be available here.

Reactive properties

Reactive properties are properties of the component option object that let us synchronize what is displayed in the template, and they change according to the operations we do with them. Any changes that are applied to reactive properties are propagated throughout the app wherever they are referenced.

In the previous example, we added the `count` reactive property to our app. To update it, we just have to update the reactive property's value itself:

```
<!DOCTYPE html>
<html lang="en">
  <head>
    <title>Vue App</title>
    <script src="https://unpkg.com/vue@next"></script>
  </head>
  <body>
    <div id="app">
      <button @click="count++">increment</button>
      count: {{ count }}
    </div>
    <script>
      const Counter = {
        data() {
          return {
            count: 0
```

```
            };
          }
        };

        Vue.createApp(Counter).mount("#app");
    </script>
  </body>
</html>
```

Here, we have the `@click="count++"` expression, which listens for clicks of the button, and we increase the count by 1 when we click the increment button. The latest value will be reflected everywhere since it is a reactive property. Vue can pick up the changes for reactive properties automatically. `@click` is shorthand for `v-on:click`.

Also, we can rewrite the expression as a method. To do that, we can write the following code:

```
<!DOCTYPE html>
<html lang="en">
  <head>
    <title>Vue App</title>
    <script src="https://unpkg.com/vue@next"></script>
  </head>
  <body>
    <div id="app">
      <button @click="increment">increment</button>
      count: {{ count }}
    </div>
    <script>
      const Counter = {
        data() {
          return {
            count: 0
          };
        },
        methods: {
          increment() {
            this.count++;
```

```
        }
      }
    };

    Vue.createApp(Counter).mount("#app");
  </script>
  </body>
</html>
```

To reference the `count` reactive property in the Vue instance object, we must reference it as a property of this. So, `this.count` in the Vue instance object is the same as `count` in the template. The `this` keyword refers to the component instance. We should remember this so that we don't run into problems later.

Also, we add the method's properties to the component object. This is a special property that is used to hold methods in our code that we can reference in other parts of the Vue instance or in our template. Like with reactive properties, methods are referenced as properties of this in the Vue instance object, and we omit this in the template.

So, when we click the button, we run the increment method in the `methods` property. When we click the button, the count value will increase by 1, and we should see that displayed in our browser's output.

Handling user input

Most apps require users to input something to forms. We can do this easily with Vue 3 with the `v-model` directive. It synchronizes the inputted value with the reactive properties that we have defined in our Vue instance.

To use it, we just add the `v-model` attribute to the input box. To do that, we can write the following code:

```
<!DOCTYPE html>
<html lang="en">
  <head>
    <title>Vue App</title>
    <script src="https://unpkg.com/vue@next"></script>
  </head>
  <body>
    <div id="app">
      <p>{{ message }}</p>
```

```
      <input v-model="message" />
    </div>
    <script>
      const App = {
        data() {
          return {
            message: "hello world."
          };
        }
      };

      Vue.createApp(App).mount("#app");
    </script>
  </body>
</html>
```

Here, we have the message reactive property, which has been initialized to the hello world. string. We can use the same value in the template by setting that as the value of the v-model directive. It will do the synchronization between the inputted value and the message reactive property so that whatever we type in will be propagated to the rest of the Vue instance.

Therefore, the hello world. string is both shown in the input box and the paragraph element. And when we enter something into the input box, it will also be shown in the paragraph element. It will update the value of the message reactive property. This is one great feature that comes with Vue 3 that we will use in many places.

Conditionals and loops

Another very useful feature of Vue 3 is that we can conditionally render content in the template. To do this, we can use the v-if directive, which lets us show something conditionally. The v-if directive puts the element in the DOM only if the condition we assign to it is true. The v-show directive shows and hides the element it is bound to with CSS, and the element is always in the DOM. If the value for it is true, then we will see it shown in the template. Otherwise, we don't see the item displayed.

It works by conditionally attaching the item to the DOM. The elements and components that are inside the element or component that has the v-if directive are only appended to the DOM when the v-if value is true. Otherwise, they won't be attached to the DOM.

For instance, let's say we have the following code:

```html
<!DOCTYPE html>
<html lang="en">
  <head>
    <title>Vue App</title>
    <script src="https://unpkg.com/vue@next"></script>
  </head>
  <body>
    <div id="app">
      <span v-if="show">hello world</span>
    </div>
    <script>
      const App = {
        data() {
          return {
            show: true
          };
        }
      };

      Vue.createApp(App).mount("#app");
    </script>
  </body>
</html>
```

Here, `'hello world'` will be shown since `show` is `true`. If we have the following code, we won't see anything displayed since the span isn't attached to the DOM:

```html
<!DOCTYPE html>
<html lang="en">
  <head>
    <title>Vue App</title>
    <script src="https://unpkg.com/vue@next"></script>
  </head>
  <body>
    <div id="app">
      <span v-if="show">hello world</span>
```

```
    </div>
    <script>
      const App = {
        data() {
          return {
            show: false
          };
        }
      };

      Vue.createApp(App).mount("#app");
    </script>
  </body>
</html>
```

To render an array of items in our template and the final output, we can use the v-for directive. We place a value that is a special JavaScript expression that lets us loop through the array. We can use the v-for directive by writing the following code:

```
<!DOCTYPE html>
<html lang="en">
  <head>
    <title>Vue App</title>
    <script src="https://unpkg.com/vue@next"></script>
  </head>
  <body>
    <div id="app">
      <ol>
        <li v-for="todo in todos" :keu="todo.id">
          {{ todo.description }}
        </li>
      </ol>
    </div>
    <script>
      const App = {
        data() {
          return {
```

```
        todos: [
            { description: "eat", id: 1 },
            { description: "drink", id: 2 },
            { description: "sleep", id: 3 }
        ]
      };
    }
  };

  Vue.createApp(App).mount("#app");
  </script>
  </body>
</html>
```

We have the v-for directive with our li element. 'todo in todos' loops through the todo array and renders the item in-between the tags. The todo variable is the individual *todos* entry that is being iterated. We access the description property so that we can show the value of the description in the list.

Once we have done that, we will see a numbered list with the todo text listed.

Template syntax

We have already used templates extensively. We mainly use interpolation to display data and some directives to render data. Also, we can use the @ or v-on directive to listen to events that are emitted, such as clicks and input value changes.

There is other useful syntax that we can use to create templates with. One of them is displaying raw HTML with our interpolated expressions. By default, Vue 3 escapes all HTML entities so that they will be displayed as-is. The v-html directive lets us display HTML code as real HTML rather than as plain text.

For example, let's write the following code:

```
<!DOCTYPE html>
<html lang="en">
  <head>
    <title>Vue App</title>
    <script src="https://unpkg.com/vue@next"></script>
  </head>
  <body>
```

```
    <div id="app">
      <span v-html="rawHtml"></span>
    </div>
    <script>
      const App = {
        data() {
          return {
            rawHtml: `<b>hello world</b>`
          };
        }
      };

      const app = Vue.createApp(App);
      app.mount("#app");
    </script>
  </body>
</html>
```

Here, we set the `rawHtml` reactive property as the value of `v-html`, so that we can see the b tag being rendered as bold text instead of the characters in raw form being rendered.

JavaScript expressions and templates

We can put any JavaScript expressions in-between the curly braces. It can only be a single expression.

For example, the following pieces of code show what's valid in-between the curly braces:

```
{{ number + 1 }}
{{ areYouSure ? 'YES' : 'NO' }}
{{ message.split('').reverse().join('') }}
```

However, we can't put any JavaScript statements in-between the curly braces. For example, we can't write `{{ var foo = 1 }}` or `{{ if (yes) { return message } }}`.

Computed properties

Computed properties are special reactive properties that are derived from other reactive properties. Computed properties are added to the computed property's objects as functions. They always return something that is derived from other reactive properties. Therefore, they must be synchronous functions.

To create a computed property, we can write the following code:

```html
<!DOCTYPE html>
<html lang="en">
  <head>
    <title>Vue App</title>
    <script src="https://unpkg.com/vue@next"></script>
  </head>
  <body>
    <div id="app">
      <p>{{message}}</p>
      <p>{{reversedMessage}}</p>
    </div>
    <script>
      const App = {
        data() {
          return {
            message: "hello world"
          };
        },
        computed: {
          reversedMessage() {
            return
              this.message.split("").reverse().join("");
          }
        }
      };

      const app = Vue.createApp(App);
      app.mount("#app");
    </script>
```

```
    </body>
</html>
```

Here, we created the `reversedMessage` computed property, which is the reverse of the `message` reactive property. We return the message with the order of the characters reversed. Whenever the `message` reactive property is updated, the `reversedMessage()` method will be run again and return the newest value. Therefore, we can see both `'hello world'` and `'dlrow olleh'` in the same template. The return values of these computed properties must have other reactive properties in them so that they will be updated when other reactive properties update.

Directives

Components may not have enough to do what we want. The main thing that is missing is the ability to manipulate the DOM and synchronize input data with reactive properties. Directives are special attributes that start with the `v-` prefix. They expect single JavaScript expressions as values. We have already seen some built-in directives such as `v-if`, `v-for`, `v-bind`, and `v-on` being used for various purposes. Directives can take arguments in addition to values.

For example, we can write `<a v-on:click="doSomething"> ... </a>` to listen to the click event on the anchor element. The `v-on` part is the directive's name. The part between the colon and the equal sign is the argument for the directive, so `click` is the directive's argument. `doSomething` is the value of the directive. It is the name of the method we want to call.

Directive arguments can be dynamic. To add dynamic arguments, we can put them between square brackets:

```
<a v-bind:[attributeName]="url"> ... </a>
```

`attributeName` is the reactive property that we want to use to set the value of the argument. It should be a string. We can also do the same thing with the `v-on` directive:

```
<a v-on:[eventName]="doSomething"> ... </a>
```

We listen to the event with the given `eventName`. `eventName` should also be a string.

Directive modifiers

Directives can take modifiers that let us change the behavior of a directive. Modifiers are special postfixes that are denoted by a dot. They can be chained to provide more changes. They indicate that a directive should be bound in some special way. For instance, if we need to listen to the `submit` event, we can add the `prevent` modifier to make it call `event.preventDefault()`, which will prevent the default submission behavior. We can do that by writing the following code:

```
<form v-on:submit.prevent="onSubmit">...</form>
```

Next, we will look at how to debug Vue 3 projects easily with the Vue.js Devtools browser extension.

Debugging with Vue.js Devtools

Right now, there is no easy way to debug our app. All we can do is add `console.log` statements to our code to look at the values. With Vue.js Devtools, we can have more visibility in our app. Vue.js Devtools is a Chrome or Firefox extension that we can use to debug our Vue.js applications. It can be used on projects that are created with Vite or created from scratch by including the `script` tag for Vue 3. We can install the extension by searching for the Vue.js Devtools extension in the respective browser's app store.

> **Important:**
>
> The URL to install the Chrome version of Vue.js Devtools is at `https://chrome.google.com/webstore/detail/vuejs-devtools/nhdogjmejiglipccpnnnanhbledajbpd`.
>
> The Firefox version of the add-on is at `https://addons.mozilla.org/en-CA/firefox/addon/vue-js-devtools/?utm_source=addons.mozilla.org&utm_medium=referral&utm_content=search`.

Once we've installed it, we should see the Vue tab in the browser's development console. With it, we can inspect the reactive properties that are loaded by Vue. If our component has a `name` property, then it will be displayed in the component tree of the application. For example, let's say we have the following code:

```
<!DOCTYPE html>
<html lang="en">
  <head>
    <title>Vue App</title>
```

```
        <script src="https://unpkg.com/vue@next"></script>
    </head>
    <body>
      <div id="app">
        <foo></foo>
      </div>
      <script>
        const App = {
          data() {
            return {};
          }
        };
        const app = Vue.createApp(App);

        app.component("foo", {
          data() {
            return {
              message: "I am foo."
            };
          },
          name: "foo",
          template: `<p>{{message}}</p>`
        });

        app.mount("#app");
      </script>
    </body>
</html>
```

Here, since we have the `name` property of the `foo` component set to `'foo'`, we will see
that listed in the component tree. Also, the `foo` component has the `message` reactive
property, so we will also see the `message` property displayed with its value. Above the
component tree, there is a search box that lets us find the reactive property with the given
name. We can also search for components with the **Find components...** input box.

The following screenshot shows us the values of Reactive properties in our Vue 3 app, within the Vue Devtools extension:

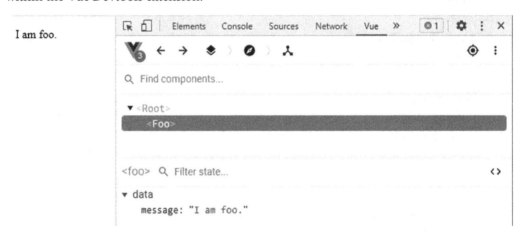

Figure 1.1 – Inspecting reactive properties with Vue Devtools

There is also the **Timeline** menu item, which we can use to inspect the events that are emitted. For example, let's say we have the following code:

```
<!DOCTYPE html>
<html lang="en">

<head>
  <title>Vue App</title>
  <script src="https://unpkg.com/vue@next"></script>
</head>

<body>
  <div id="app">
    <button @click="count++">increment</button>
    count: {{ count }}
  </div>
  <script>
    const Counter = {
      data() {
        return {
          count: 0
        };
```

```
      }
    };

    Vue.createApp(Counter).mount("#app");
  </script>
</body>

</html>
```

When we click on the **Increment** button, we will see the mouse events logged in the **Timelines** section. The time the event is triggered will also be logged.

In the **Global Settings** section, we can see the **Normalize Component Names** setting, which lets us change how the component names are displayed. We can display the original name in Pascal case or Kebab case. The **Theme** option lets us change the theme color of the Vue tab.

Summary

In this chapter, we learned that Vue 3 is a component-based framework, and we looked at the different parts of the component. One important part we covered was reactive properties. They are properties of the component that we can change to update all the parts of the app that reference the reactive property. These properties can be watched manually, and any value changes are also picked up automatically by Vue 3 so that any parts of the app that reference the reactive property are updated automatically. Components are composed in a way that they can be reused whenever possible.

We then moved on to understand the role of a template in every component. Templates are also a part of every component. They must all render something onto the screen. Templates can have HTML elements, other components, and directives that modify how the elements and components in the template are rendered. Templates can have JavaScript expressions in them so that we can do things such as react to events. Then, we looked at the importance of computed properties in a component. Computed properties are special reactive properties that depend on other reactive properties. They are functions that are synchronous and return a value based on combining other reactive properties.

Another important point that we looked at is the v-model directive, which is built into Vue 3. Vue 3 provides the v-model directive so that we can bind reactive properties to form control values. Directives are special Vue code that let us change how DOM elements are rendered. Vue 3 provides many built-in directives to do things such as render elements from an array, bind form control values to reactive properties, and more.

In the last section, we looked at how to use Vue.js Devtools to make debugging easier. It is a browser extension available for Chromium browsers and Firefox that lets us watch the reactive property values of components and see what components are rendered. It will also log any events that are triggered by the elements in a component.

In the next chapter, we will look at how to build a simple GitHub app that makes HTTP requests.

2
Building a Vue 3 Progressive Web App

In this chapter, we will look at how to create a GitHub **progressive web app** (PWA) with Vue 3. As we build the project, we will examine the inner workings of a Vue app by looking at the basic building blocks in depth. We will create Vue apps with components and, as we create them, we will look at the parts that make up a component and how they work.

We will also use more advanced features such as directives, when we need to do so. Directives let us manipulate the **Document Object Model** (DOM) without cluttering up a component's code. They provide us with a clean way to access DOM elements and work with them in a reusable way. This helps make testing easier and helps us to modularize our code.

Vue 3 comes with many built-in directives that we will use. In the previous chapter, we saw a brief overview of these. In this chapter, we will go into more detail to see how they work. These directives provide easy-to-use abstractions to make many things easier for us and are a basic feature of Vue 3 that we can't live without.

We will use components to display the data we want, which will take in inputs via props so that we can get the proper data and display it. In each component, we will add our own methods and make use of some component lifecycle methods. To reduce repetition of code, we use mixins to abstract out commonly used features in components and incorporate them into our components.

In this chapter, we will look at the following topics:

- Basic theory on components and PWAs

- Introducing the GitHub portfolio app

- Creating the PWA

- Serving the PWA

Technical requirements

The code for this chapter can be found at `https://github.com/PacktPublishing/-Vue.js-3-By-Example/tree/master/Chapter02`.

Basic theory on components and PWAs

Before we begin with building our Vue app, let's first get familiar with components and PWA. Vue 3 lets us build frontend web apps with components. With them, we can divide our app into small, reusable parts that are composed together to make a big app. This composition is done by nesting. To make different parts of the app compose together, we can pass data between them. Components can be taken from libraries and can also be created by us.

A component consists of several parts; it includes a template, a script, and styles. The template is what is rendered on the screen. It has **HyperText Markup Language** (**HTML**) elements, directives, and components. Components and HTML elements can have props and event listeners added to them. Props are used to pass data from a parent component to a child component.

Event listeners let us listen to events emitted from a child component to a parent component. Events may be emitted with a payload, with data included in it. This enables us to have child component-to-parent component communication. With both things put together, we have a complete system to communicate between parent and child components.

Any non-trivial app will have multiple components that need to communicate with each other.

PWAs are special web apps that can be installed on the user's computer, and the browser manages these installed apps. They differ from regular web apps as they let us access some computer hardware natively. When we visit a PWA in our browsers, we can choose to install the PWA and can then reach our app from the app store.

PWAs don't require special bundling or distribution procedures. This means they are deployed just like any other web app to a server. Many modern browsers—such as Mozilla Firefox, Google Chrome, Apple Safari, and Microsoft Edge—support PWAs. This means that we can install the apps with them.

Special characteristics of PWAs include the ability to work for every user, regardless of browser choice. They are also responsive, which means they work on any device, such as desktop, laptop, tablet, or mobile devices. Initial loading is also fast since they are supposed to be cached on first load.

They are also supposed to work regardless of whether there's connectivity to the internet. Service workers run in the background to let us use PWAs offline or on low-quality networks. This is also another benefit of the caching available to PWAs.

Even though PWAs are run from the browser, they act like apps. They have app-like style interactions and navigation. Whatever is displayed is also always up to date, since the service worker runs in the background to update the data.

Security is a further important benefit of PWAs. They can only be served over **HTTP Secure (HTTPS)**, so outsiders can't snoop on the connection. This way, we know the connection hasn't been tampered with.

Push notifications are also available with PWAs so that they can engage with the user and notify them of updates.

They can also be linked from a **Uniform Resource Locator (URL)**, and a PWA doesn't require an installation process before we can use it—installation is strictly optional. When we install it, it provides a home screen icon on our browser so that we can click on it and start using it.

Vue 3 has a `@vue/cli-plugin-pwa` plugin to let us add PWA abilities into our Vue 3 project without doing any manual configuration. We just run one command and have all the files and configuration added for us automatically. With this plugin, we can develop our PWA with Vue 3, and the included service worker will run in production. Now that we have this out of the way, we are going to look at how to create reusable components.

Introducing the GitHub portfolio app

The main project of this chapter is a GitHub portfolio app. It is a PWA, which means it has all the features listed in the *Basic theory on components and PWAs* section of this chapter. These features are provided automatically by the `@vue/cli-plugin-pwa` plugin. We can add the code we need, to add the service workers and any other required configuration with one command. This way, we don't have to configure everything all by ourselves from scratch when we create our Vue project.

To get started with our app project, we will create it using Vite. We go into the folder where we want our project to be, and then run Vite to create the Vue 3 app project. To do this, we run the following commands with **Node Package Manager** (**npm**):

1. The first command, shown in the following code snippet, runs npm to install the Vue **command-line interface** (**CLI**) globally:

    ```
    npm install -g @vue/cli@next
    ```

2. We run the Vue CLI to create our Vue 3 project. Our project folder name is `vue-example-ch2-github-app`. The following command is needed to create the project folder with all the files and settings added so that we don't have to add them ourselves. This command goes to the project folder we just created and chooses the Vue 3 project when asked:

    ```
    npm vue create vue-example-ch2-github-app
    ```

3. Then, we run the following command to run the development server so that we can see the project in the browser and refresh the app preview when we write our code:

    ```
    npm run serve
    ```

Alternatively, we can run the following commands with **Yet Another Resource Negotiator** (**YARN**):

1. We run `yarn global add` to install the Vue CLI globally, as follows:

    ```
    yarn global add @vue/cli@next
    ```

2. To create the Vue 3 project, we run the following command and choose the Vue 3 project when asked:

    ```
    yarn create vue-example-ch2-github-app
    ```

3. Then, we run the following command to run the development server so that we can see the project in the browser and refresh the app preview when we write our code:

```
yarn serve
```

All the preceding commands are the same, as in they both create the project the same way; it's just a matter of which package manager we want to use to create our Vue 3 project. At this point, the project folder will have the required files for our Vue 3 project.

Our GitHub portfolio app is a progressive web app, and we can create this app easily with an existing Vue CLI plugin. Once we have created the project, we can start creating our Vue 3 PWA.

Creating the PWA

First, we need an easy way to access GitHub data via its **Representational State Transfer (REST) application programming interface (API)**. Fortunately, an developer named *Octokit* has made a JavaScript client that lets us access the GitHub REST API with an access token that we create. We just need to import the package from the **content distribution network (CDN)** that it is served from to get access to the GitHub REST API from our browser. It also has a Node package that we can install and import. However, the Node package only supports Node.js apps, so it can't be used in our Vue 3 app.

Vue 3 is a client-side web framework, which means that it mainly runs on the browser. We shouldn't confuse packages that only run on Node with packages that support the browser, otherwise we will get errors when we use unsupported packages in the browser.

To get started, we make a few changes to the existing files. First, we remove the styling code from `index.css`. We are focused on the functionality of our app for this project and not so much on the styles. Also, we rename the title tag's inner text to `GitHub App` in the `index.html` file.

Then, to make our built app a PWA, we must run another command to add the service worker, to incorporate things such as hardware access support, installation, and support for offline usage. To do this, we use the `@vue/cli-plugin-pwa` plugin. We can add this by running the following command:

```
vue add pwa
```

This will add all the files and configurations we need to incorporate to make our Vue 3 project a PWA project.

Vue CLI creates a Vue project that uses single-file components and uses **ECMAScript 6 (ES6)** modules for most of our app. When we build the project, these are bundled together into files that are served on the web server and run on the browser. A project created with Vue CLI consists of `main.js` as its entry point, which runs all the code that is needed to create our Vue app.

Our `main.js` file should contain the following code:

```
import { createApp } from 'vue'
import App from './App.vue'
import './registerServiceWorker'


createApp(App).mount('#app')
```

This file is located at the root of the `src` folder, and Vue 3 will automatically run this when the app first loads or refreshes. The `createApp` function will create the Vue 3 app by passing in the entry-point component. The entry-point component is the component that is first run when we first load our app. In our project, we imported `App` and passed it into `createApp`.

Also, the `index.css` file is imported from the same folder. This has the global styles of our app, which is optional, so if we don't want any global styles, we can omit it. The `registerServiceWorker.js` file is then imported. An import with the filename only means that the code in the file is run directly, rather than us importing anything from the module.

The `registerServiceWorker.js` file should contain the following code:

```
/* eslint-disable no-console */

import { register } from 'register-service-worker'

if (process.env.NODE_ENV === 'production') {
...
    offline () {
      console.log('No internet connection found. App is running
          in offline mode.')
    },
    error (error) {
      console.error('Error during service worker
          registration:', error)
```

```
        }
    })
}
```

This is what we created when we ran vue add pwa. We call the register function to register the service worker if the app is in production mode. When we run the npm run build command, the service worker will be created, and we can use the service worker that is created to let users access features—such as caching and hardware access—from the built code that we serve. The service worker is only created in production mode since we don't want anything to be cached in the development environment. We always want to see the latest data displayed so that we can create code and debug it without being confused by the caching.

One more thing we need to do is to remove the HelloWorld.vue component from the src/components folder, since we don't need this in our app. We will also remove any reference to the HelloWorld component in App.vue later.

Now that we have made the edits to the existing code files, we can create the new files. To do this, we carry out the following steps:

1. In the components folder, we add a repo folder; and in the repo folder, we add an issue folder. In the repo folder, we add the Issues.vue component file.

2. In the components/repo/issue folder, we add the Comments.vue file. Issues.vue is used to display the issues of a GitHub code repository. Comments.vue is used to display the comments that are added to an issue of the code repository.

3. In the components folder itself, we add the GitHubTokenForm.vue file to let us enter and store the GitHub token.

4. We also add the Repos.vue file to the same folder to display the code repositories of the user that the GitHub access token refers to. Then, finally, we add the User.vue file to the components folder to let us display the user information.

5. Create a mixins folder in the src folder to add a mixin, to let us create the Octokit GitHub client with the GitHub access token.

We add the octokitMixin.js file to the mixins folder to add the empty mixin. Now, we leave them all empty, as we are ready to add the files.

Creating the GitHub client for our app

We start the project by creating the GitHub `Client` object that we will use throughout the app.

First, in the `src/mixins/octokitMixin.js` file, we add the following code:

```
import { Octokit } from "https://cdn.skypack.dev/@octokit/
rest";

export const octokitMixin = {
  methods: {
    createOctokitClient() {
      return new Octokit({
        auth: localStorage.getItem("github-token"),
      });
    }
  }
}
```

The preceding file is a mixin, which is an object that we merge into components so that we can use it correctly in our components. Mixins have the same structure as components. The `methods` property is added so that we can create methods that we incorporate into components. To avoid naming conflicts, we should avoid naming any method with the name `createOctokitClient` in our components, otherwise we may get errors or behaviors that we don't expect. The `createOctokitClient()` method uses the Octokit client to create the client by getting the `github-token` local storage item and then setting that as the `auth` property. The `auth` property is our GitHub access token.

The `Octokit` constructor comes from the `octokit-rest.min.js` file that we add from `https://github.com/octokit/rest.js/releases?after=v17.1.0`. We find the `v16.43.1` heading, click on **Assets**, download the `octokit-rest.min.js` file, and add it to the `public` folder. Then, in `public/index.html`, we add a `script` tag to reference the file. We should have the following code in the `index.html` file:

```
<!DOCTYPE html>
<html lang="en">
  <head>
    <meta charset="utf-8">
    <meta http-equiv="X-UA-Compatible" content="IE=edge">
```

```
    <meta name="viewport" content="width=device-
      width,initial-scale=1.0">
    <link rel="icon" href="<%= BASE_URL %>favicon.ico">
    <title><%= htmlWebpackPlugin.options.title %></title>
    <script src="<%= BASE_URL %>octokit-rest.min.js">
      </script>
  </head>
  <body>
    <noscript>
      <strong>We're sorry but <%= htmlWebpackPlugin.
          options.title %> doesn't work properly without
          JavaScript enabled. Please enable it to
          continue.</strong>
    </noscript>
    <div id="app"></div>
    <!-- built files will be auto injected -->
  </body>
</html>
```

Adding a display for issues and comments

Then, in the `src/components/repo/issue/Comments.vue` file, we add the following code:

```
<template>
  <div>
    <div v-if="comments.length > 0">
      <h4>Comments</h4>
      <div v-for="c of comments" :key="c.id">
        {{c.user && c.user.login}} - {{c.body}}
      </div>
    </div>
  </div>
...
      repo,
      issue_number: issueNumber,
    });
```

```
      this.comments = comments;
    }
  },
  watch: {
    owner() {
      this.getIssueComments();
    },
    repo() {
      this.getIssueComments();
    },
    issueNumber() {
      this.getIssueComments();
    }
  }
};
</script>
```

In this component, we have a `template` section and a `script` section. The `script` section has our logic to get the comments from an issue. The `name` property has the name of our component. We reference our component with this name in our other components, if needed. The `props` property has the props that the component accepts, as shown in the following code snippet:

```
props: {
  owner: {
    type: String,
    required: true,
  },
  repo: {
    type: String,
    required: true,
  },
  issueNumber: {
    type: Number,
    required: true,
  },
},
```

The component takes the `owner`, `repo`, and `issueNumber` props. We use an object to define the props so that we can validate the type easily with the `type` property. The type for `owner` and `repo` has the value `String`, so they must be strings. The `issueNumber` property has the type value set to `Number`, so it must be a number.

The `required` property is set to `true`, which means that the `prop` must be set when we use the `Comments` component in another component.

The `data()` method is used to return an object that has the initial values of reactive properties. The `comments` reactive property is set to an empty array as its initial value.

The `mixins` property lets us set the mixins that we want to incorporate into our app. Since `octokitMixin` has a `methods` property, whatever is inside will be added into the `methods` property of our component so that we can call the components directly, as we will do in the `methods` property of this component.

We incorporate our mixin into our component object, as follows:

```
mixins: [octokitMixin],
```

In the `methods` property, we have one method in our `Comments` component. We use the `getIssueComments()` method to get the comments of an issue. The code for this is shown in the following snippet:

```
...
methods: {
  ...
  async getIssueComments(owner, repo, issueNumber) {
    if (
      typeof owner !== "string" ||
      typeof repo !== "string" ||
      typeof issueNumber !== "number"
    ) {
      return;
    }
    const octokit = this.createOctokitClient();
    const { data: comments } = await
      octokit.issues.listComments({
      owner,
      repo,
      issue_number: issueNumber,
```

```
    });
    this.comments = comments;
  },
  ...
 }
 ...
}
```

We need the `owner`, `repo`, and `issueNumber` properties. The `owner` parameter is the username of the user who owns the repository, the `repo` parameter is the repository name, and the `issueNumber` parameter is the issue number of the issue.

We check for the types of each to make sure that they are what we expect before we make a request to get the issue, with the `octokit.issue.listComments()` method. The Octokit client is created by the `createOctokitClient()` method of our mixin. The `listComments()` method returns a promise that resolves the issue with the comments data.

After that, we have the `watch` property to add our watchers. The keys of the properties are the names of the props that we are watching. Each object has an `immediate` property, which makes the watchers start watching as soon as the component loads. The `handler` methods have the handlers that are run when the prop value changes or when the component loads, since we have the `immediate` property set to `true`.

We pass in the required values from the properties of this, along with `val` to call the `getIssueComments()` method. The `val` parameter has the latest value of whatever prop that we are watching. This way, we always get the latest comments if we have values of all the props set.

In the template, we load the comments by referencing the `comments` reactive property. The values are set by the `getIssueComments()` method that is run in the watcher. With the `v-for` directive, we loop through each item and render the values. The `c.user.login` property has the username of the user who posted the comment, and `c.body` has the body of the comment.

Next, we add the following code to the `src/components/Issues.vue` file:

```
...
<script>
import { octokitMixin } from "../../mixins/octokitMixin";
import IssueComments from "./issue/Comments.vue";
```

```
export default {
  name: "RepoIssues",
  components: {
    IssueComments,
  },
  props: {
    owner: {
      type: String,
      required: true,
    },
    repo: {
      type: String,
      required: true,
    },
  },
  mixins: [octokitMixin],
  ...
};
</script>
```

The preceding code adds a component for displaying the issues. We have similar code in the Comments.vue component. We use the same octokitMixin mixin to incorporate the createOctokitClient() method from the mixin.

The difference is that we have the getRepoIssues() method to get the issues for a given GitHub repository instead of the comments of a given issue, and we have two props instead of three. The owner and repo props are both strings, and we make them required and validate their types in the same way.

In the data() method, we have the issues array, which is set when we call getRepoIssues. This is shown in the following code snippet:

src/components/Issues.vue

```
data() {
  return {
    issues: [],
```

```
        showIssues: false,
      };
    },
```

The `octokit.issues.listForRepo()` method returns a promise that resolves the issues for a given repository. The `showIssue` reactive property lets us toggle whether to show the issues or not.

We also have methods to get the GitHub issues, as illustrated in the following code snippet:

src/components/Issues.vue

```
    methods: {
      async getRepoIssues(owner, repo) {
        const octokit = this.createOctokitClient();
        const { data: issues } = await
          octokit.issues.listForRepo({
          owner,
          repo,
        });
        this.issues = issues;
      },
    },
```

The `showIssues` reactive property is controlled by the **Show issues** button. We use the `v-if` directive to show the issues when the `showIssues` reactive property is `true`. The outer `div` tag is used for checking the length property of issues so that we only show the **Show issues** button and the issues list when the length is greater than `0`.

The method is triggered by the watchers, as follows:

src/components/Issues.vue

```
    watch: {
      owner: {
        handler(val) {
          this.getRepoIssues(val, this.repo);
        },
      },
```

```
    repo: {
      handler(val) {
        this.getRepoIssues(this.owner, val);
      },
    },
  },
  created () {
    this.getRepoIssues(this.owner, this.repo);
  }
```

In the `components` property, we put the `IssueComments` component we imported (the one we created earlier) into our component object. If we put the component in the `components` property, it is then registered in the component and we can use it in the template.

Next, we add the template into the file, as follows:

src/components/Issues.vue

```html
<template>
  <div v-if="issues.length > 0">
    <button @click="showIssues = !showIssues">{{showIssues
      ? 'Hide' : 'Show'}} issues</button>
    <div v-if="showIssues">
      <div v-for="i of issues" :key="i.id">
        <h3>{{i.title}}</h3>
        <a :href="i.url">Go to issue</a>
        <IssueComments :owner="owner" :repo="repo"
          :issueNumber="i.number" />
      </div>
    </div>
  </div>
</template>
```

When we use the `v-for` directive, we need to include the `key` prop so that the entries are displayed correctly, for Vue 3 to keep track of them. The value of `key` must be a unique ID. We reference the `IssueComments` component we registered in the template and pass in the `props` to it. The `:` symbol is short for the `v-bind` directive, to indicate that we are passing props to a component instead of setting an attribute.

Letting users access GitHub data with a GitHub token

Next, we work on the `src/components/GitHubTokenForm.vue` file, as follows:

```
<template>
  <form @submit.prevent="saveToken">
    <div>
      <label for="githubToken">Github Token</label>
      <br />
      <input id="githubToken" v-model="githubToken" />
    </div>
    <div>
      <input type="submit" value="Save token" />
      <button type="button" @click="clearToken">Clear token
        </button>
...
    clearToken() {
      localStorage.clear();
    },
  },
};
</script>
```

We have a form that has an input to let us enter the GitHub access token. This way, we can save it when we submit the form. Also, we have the input, with type `submit`. The `value` attribute of it is shown as the text for the **Submit** button. We also have a button that lets us clear the token. The `@submit.prevent` directive lets us run the `saveToken` submit handler and call `event.preventDefault()` at the same time. The `@` symbol is short for the `v-on` directive, which listens to the submit event emitted by the form.

The text input has a `v-model` directive to bind the input value to the `githubToken` reactive property. To make our input accessible for screen readers, we have a label with a `for` attribute that references the ID of the input. The text between the tags is displayed in the label.

Once the form is submitted, the `saveToken()` method runs to save the inputted value to local storage with the `github-token` string as the key. The `created()` method is a lifecycle hook that lets us get the value from local storage. The item with the `github-token` key is accessed to get the saved token.

The `clearToken()` method clears the token and is run when we click on the **Clear token** button.

Next, we add the following code to the `src/components/Repos.vue` component:

```
<template>
  <div>
    <h1>Repos</h1>
    <div v-for="r of repos" :key="r.id">
      <h2>{{r.owner.login}}/{{r.name}}</h2>
      <Issues :owner="r.owner.login" :repo="r.name" />
    </div>
  </div>
</template>

<script>
import Issues from "./repo/Issues.vue";
import { octokitMixin } from "../mixins/octokitMixin";

export default {
  name: "Repos",
  components: {
    Issues,
  },
  data() {
    return {
      repos: [],
    };
  },
  mixins: [octokitMixin],
  async mounted() {
    const octokit = this.createOctokitClient();
    const { data: repos } = await
      octokit.request("/user/repos");
```

```
    this.repos = repos;
  },
};
</script>
```

We make a request to the `/user/repos` endpoint of the GitHub REST API with the `octokit.request()` method. Once again, the `octokit` object is created with the same mixin that we used before. We register the `Issues` component so that we can use it to display the issues of the code repository. We loop through the `repos` reactive property, which is assigned the values from the `octokit.request()` method.

The data is rendered in the template. The `r.owner.login` property has the username of the owner of the GitHub repository, and the `r.name` property has the repository name. We pass both values as props to the `Issues` component so that the `Issues` component loads the issues of the given repository.

Similarly, in the `src/components/User.vue` file, we write the following code:

```
<template>
  <div>
    <h1>User Info</h1>
    <ul>
      <li>
        <img :src="userData.avatar_url" id="avatar" />
      </li>
      <li>username: {{userData.login}}</li>
      <li>followers: {{userData.followers}}</li>
      <li>plan: {{userData.pla && userData.plan.name}}</li>
    </ul>
  </div>
...
    const { data: userData } = await
      octokit.request("/user");
    this.userData = userData;
  },
};
</script>

<style scoped>
```

```
#avatar {
  width: 50px;
  height: 50px;
}
</style>
```

The scoped keyword means the styles are only applied to the current component.

This component is used to display the user information that we can access from the GitHub access token. We use the same mixin to create the octokit object for the Octokit client. The request() method is called to get the user data by making a request to the user endpoint.

Then, in the template, we show the user data by using the avatar_url property. The username.login property has the username of the owner of the token, the userData.followers property has the number of followers of the user, and the userData.plan.name property has the plan name.

Then, finally, to put the whole app together, we use the GitHubTokenForm, User, and Repo components in the App.vue component. The App.vue component is the root component that is loaded when we load the app.

In src/App.vue file, we write the following code:

```
<template>
  <div>
    <h1>Github App</h1>
    <GitHubTokenForm />
    <User />
    <Repos />
  </div>
</template>

<script>
import GitHubTokenForm from "./components/GitHubTokenForm.vue";
import Repos from "./components/Repos.vue";
import User from "./components/User.vue";

export default {
  name: "App",
```

```
components: {
  GitHubTokenForm,
  Repos,
  User,
  },
};
</script>
```

We register all three components by putting them in the `components` property to register them. Then, we use all of them in the template.

Now, we should see the following screen:

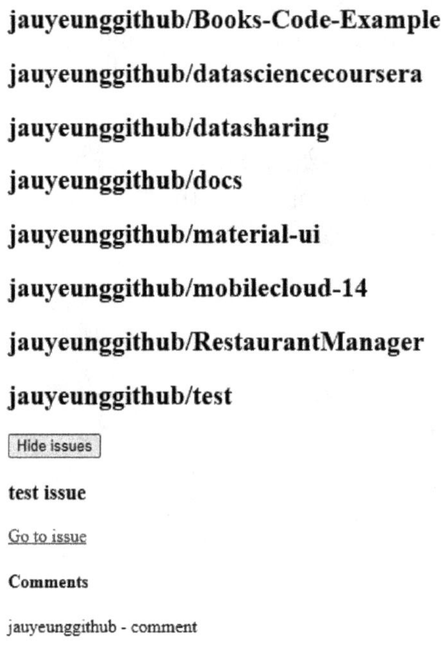

jauyeunggithub/Books-Code-Example

jauyeunggithub/datasciencecoursera

jauyeunggithub/datasharing

jauyeunggithub/docs

jauyeunggithub/material-ui

jauyeunggithub/mobilecloud-14

jauyeunggithub/RestaurantManager

jauyeunggithub/test

Hide issues

test issue

Go to issue

Comments

jauyeunggithub - comment

Figure 2.1 – List of repositories

We see a list of repositories displayed, and if there are any issues recorded for them, we see the **Show issues** button, which lets us see any issues for the given repository. This can be seen in the following screenshot:

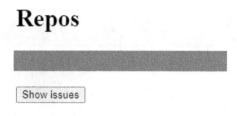

Figure 2.2 – Show issues button

We can click **Hide issues** to hide them. If there are any comments, then we should see them below the issues.

Serving the PWA

Now that we have built the app, we can serve it so that we can install it in our browser. Let's begin, as follows:

1. To build the app, we run the following command:

   ```
   npm run build
   ```

2. We can use the `browser-sync` package, which we install by running the following command:

   ```
   npm install -g browser-sync
   ```

   The preceding command will install a web server.

3. We can go into the `dist` folder, which has the built files, and run `browser-sync` to serve the PWA.

4. Now, to run the app, we need to get the GitHub authentication token from our GitHub account. If you don't have a GitHub account, then you will have to sign up for one.

5. Once we have created an account, then we can get the token. To get the token, log in to your GitHub account.

6. Go to `https://github.com/settings/tokens`.

7. Once the page is loaded, click on the **Personal access tokens** link.

8. Click **Generate new token** to generate a token. Once it's created, copy the token down somewhere so that we can use it by entering it in our app.

    We should see something like this:

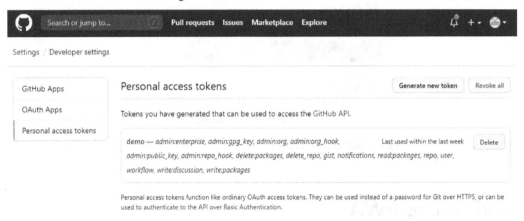

Figure 2.3 – The screen for getting the token

9. Once you have the token, go back to the app we created, which is loaded in the browser.

10. Enter the token into the **GitHub Token** input, click **Save token**, and then refresh the page. If there are any repositories and associated issues and comments, then they should show in the page.

11. Once we are in the browser, we should see a plus (+) sign on the right side of the URL bar. This button lets us install the PWA.

12. Once we install it, we should see it on the home screen. We can go to the chrome://apps URL to see the app we just installed, as shown in the following screenshot:

Figure 2.4 – The GitHub repository listing in our PWA

13. If you're using Chrome or any other Chromium browser such as Edge, you can press *F12* to open the developer console.

14. Click on the **Application** tab and then the **Service Workers** link on the left side to let us test the service worker, as illustrated in the following screenshot:

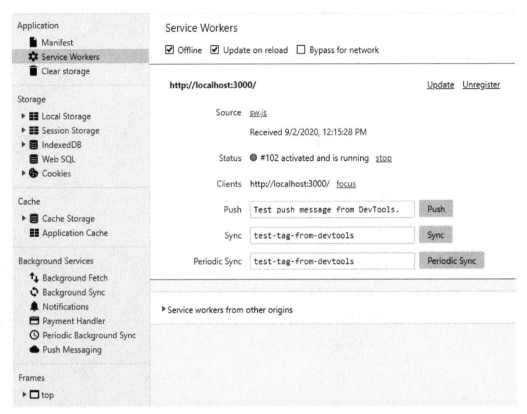

Figure 2.5 – The Service Workers section of the Application tab

15. We can check the **Offline** checkbox to simulate how it acts when it is offline. Checking the **Update on reload** will reload the app with the latest data fetched when we refresh the page. The URL should be the same as the one your app is running on. This is the service worker that is registered by our GitHub PWA.

16. The **Unregister** link will unregister the service worker. It should be re-registered when we run our app again.

We are now done with creating our progressive web app with Vue 3. We can install it with browsers and then use it like any other app on our device.

Summary

By building a GitHub PWA, we learned how to create components that can be reused. We also looked at how to add props to let us pass data from a parent component to a child component. In the child component, we validate the props by checking the data type and specifying whether a prop is required. This way, we can easily see when a prop has a value that is unexpected.

We also looked at how to use watchers to watch for changes with reactive property values. Watchers can be added to watch for changes in any reactive property. We can watch the data that is being changed locally, and also the value of props. They are both reactive, so they will both trigger the watcher methods. We can run asynchronous code within a watcher, which is something that can't be done with computed properties.

Also, we had a look at lifecycle hooks of components. Each component also has its own lifecycle hooks. We can add our own code to the lifecycle methods, to run code when we want to run them. There are lifecycle hooks for all parts of a component lifecycle, including the beginning stage when it is loaded, through to when it is updated and destroyed.

Finally, we learned how to convert our Vue 3 web app into a PWA with a command-line plugin. We can add a plugin to our Vue project to create a PWA. With it, a service worker will be registered in our app to handle different connection types and caching.

In the next chapter, we will create a slider puzzle with Vue 3, with automated tests to test each part of our app.

3
Building a Slider Puzzle Game with Tests

In the previous chapter, we created a simple GitHub app with Vue that had some components added to it. In this chapter, we will build a simple slider puzzle game. The game's goal is to rearrange parts of a picture till it looks like what we expect. It will have a timer to calculate the elapsed time and will display it on the screen. Once we rearrange the image's parts correctly, we will see a **'You Win'** message, and the elapsed time will be recorded in local storage if it is in the top 10 fastest times. We have multiple puzzles that we can choose from so that we can have more variety in our game. This makes it more interesting than just having one puzzle.

To build the apps, we will build components with computed properties and timers to calculate the elapsed time. Also, some components will get and set data from local storage. Whenever we get data from local storage, the results will be displayed. We will use local storage to store the fastest times. Local storage can only store strings, so we will convert the result into a string and store it.

We will use a timer to time when the player wins the game, and we will use computed properties to determine when the player wins the game. Also, to make sure that our game works like it should, we will add unit tests for each part to test each component automatically.

In this chapter, we will dig deeper into components and cover the following topics:

- Understanding the basics of components and mixins
- Setting up our Vue project
- Creating the components for shuffling pictures
- Letting users rearrange the slides
- Calculating the score based on timing
- Unit testing with Jest

Technical requirements

The source code for this chapter is located at `https://github.com/PacktPublishing/-Vue.js-3-By-Example/tree/master/Chapter03`.

Understanding the basics of components and mixins

There is more to components than what we did in *Chapter 2*, *Building a Vue 3 Progressive Web App*, to create the GitHub progress web app. These components were the most basic parts. We will use timers with our components, rather than just having components that get data and display it. Also, we will look at when and how to use computed properties so that we can create reactive properties that have values that are derived from other reactive properties. This saves us from creating extra methods that we don't need or using directives unnecessarily.

Furthermore, we will look at how to use computed properties to return values that are derived from other reactive properties. **Computed properties** are methods that return values that are derived from one or more other reactive properties. They are reactive properties themselves. The most common usage for them is getters. However, computed properties can have both getters and setters. Their return values are cached so that they don't run until one or more reactive properties have their values updated. They are useful for replacing complex template expressions and methods in an efficient manner.

Another thing that components can do is emit custom events. An event can contain one or more payloads emitted with the event. They have their own event name, and we can listen to the events by listening to the event with the v-on directive. We can get the emitted payload with the $event variable or the parameters of the event handler methods.

Another important part of a Vue 3 app is **tests**. When we mention tests, they are usually automated tests. Tests come in many forms and are useful for catching various kinds of bugs. They are often used for catching regressions, which are bugs that are created after we change the code that is already part of our app. We can check for regressions with a few kinds of tests. The smallest tests we can create are **unit tests**, which test a component and its parts in isolation. It works by mounting our component in a test environment. Any dependencies that prevent our tests from running in isolation are mocked so that we can run our tests in isolation. This way, we can run our tests in any environment and in any order.

Each test is independent, so we shouldn't have any issues running them anywhere, even without an internet connection. This is important because they are supposed to be portable. Also, external resources such as API data and timers are very volatile. They are also asynchronous, which makes them hard to test. Therefore, we must make sure that we don't need them for our tests since we want consistency in the results.

Vue comes with support for JavaScript test frameworks such as **Jest** and **Mocha**. This is one of the great benefits of using the Vue CLI to create our Vue project. We don't have to create all the scaffolding of the test code ourselves.

Another kind of test is *end-to-end* tests. These tests simulate how a user would use our app. We usually have an environment that is created from scratch and then taken down to run these tests. This is because we want fresh data in our tests at all times. The tests must be able to run in a consistent manner. We need consistent data for this to be done if we are going to use the app like the user.

In this chapter, we will mainly look at unit tests for our frontend app. They can provide DOM interactions like we do with end-to-end tests, but they are faster and a lot smaller. They also run a lot faster since we don't have to create a clean environment each time a test is run. The environment's creation and user interaction tests will always be slower than unit tests. Therefore, we should have many unit tests and a few end-to-end tests for testing the most critical parts of our app.

Setting up the Vue project

Now that we've learned the basics about computed properties and getters and setters, we are ready to look deeper at the component parts that we will need and create the project.

To create the project, we use the Vue CLI again. This time, instead of selecting the default options, we must choose a few options. But before we do that, we will create a project folder called `vue-example-ch3-slider-puzzle`. Then, we must go into the folder and run the following commands with npm:

1. First, we must install the Vue CLI globally so that we can create and run our project with it:

    ```
    npm install -g @vue/cli@next
    ```

2. Now, we can go into our project folder and run the following command to create our project:

    ```
    vue create .
    ```

Equivalently, we can run the following commands with Yarn:

1. First, we must install the Vue CLI globally so that we can create and run our project with it:

    ```
    yarn global add @vue/cli@next
    ```

2. Then, we can go into our project folder and run the following command to create our project:

    ```
    yarn create .
    ```

In either case, we should see the Vue CLI command-line program with instructions on how to choose the items. If we are asked if we want to create the project in the current folder, we can type *Y* and press *Enter* to do so. Then, we should see the project types that we can create our project with. We should choose `Manually select features`, and then `Vue 3` to create a Vue 3 project:

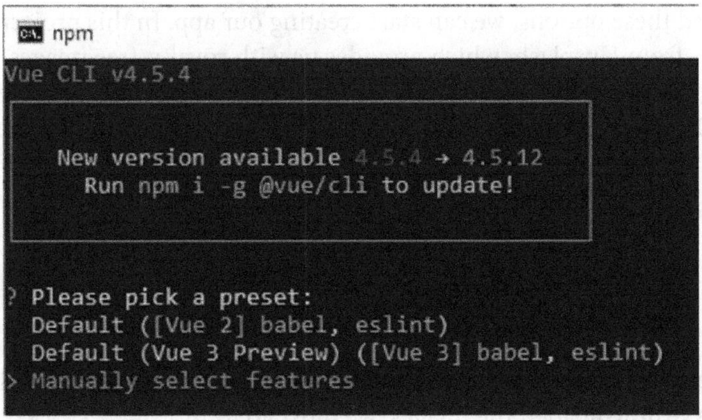

Figure 3.1 – Selecting the project type to create in the Vue CLI wizard

On the next screen, we should see what we can add to the project. Choose Unit
Testing, and then you need to choose Testing with Jest so that we can add tests
to our app.

This project will come with tests for many components once we've finished writing
the code:

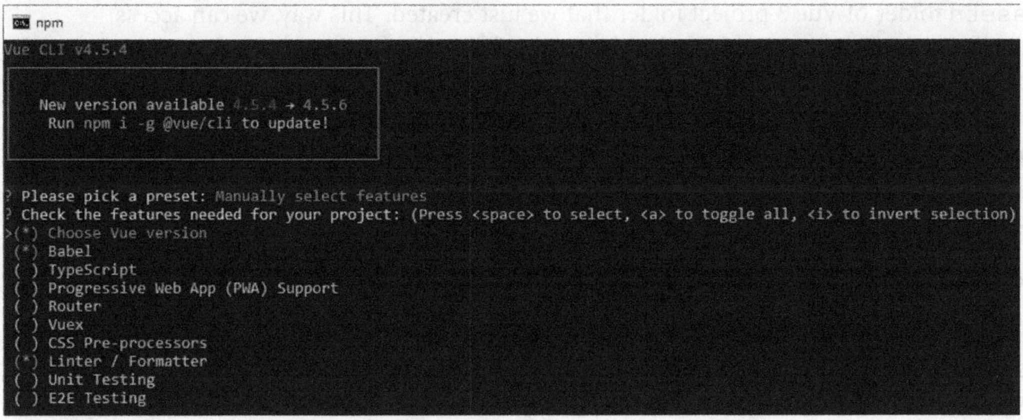

Figure 3.2 – The options we should choose for this project

Once we let the Vue CLI finish creating the project, we should see the code files in the
src folder. The tests should be in the tests/unit folder. The Vue CLI saved us lots of
effort from creating the test code all by ourselves. It comes with an example test that we
can expand from.

Once we've picked these options, we can start creating our app. In this project, we will get some pictures from Unsplash, which provides us with royalty-free images. Then, we will get the images and cut them up into nine pieces so that we can display them in the `slider puzzle` component. We need both the whole image and the cut pieces. For this example, we will get the images from the following links:

- `https://unsplash.com/photos/EfhCUc_fjrU`
- `https://unsplash.com/photos/CTvtrspsPQs`
- `https://unsplash.com/photos/XoCyW2JVmiE`

When we go to each page, we must click the **Download** button to download the images. Once we've downloaded the images, we must go to `https://www.imgonline.com.ua/eng/cut-photo-into-pieces.php` to cut the images into nine pieces automatically.

In *section 1*, we select our image file. In *section 2*, we set both **Parts in width** and **Parts in height** to 3. This way, we can divide our image into nine pieces. Once we have done that, we can download the ZIP file that is generated and then extract all the images into a folder. This should be repeated for each image.

Once we have all the whole and cut image pieces, we should put them all into the `src/assets` folder of Vue 3 project folder that we just created. This way, we can access the images from our app and display them. The first image shows a pink flower, so the whole image is named `pink.jpg` and the cut images are in the `cut-pink` folder. The filenames that are generated for the cut images remain unchanged. The second image is a purple flower, so the whole image is named `purple.jpg` and the cut image folder is named `cut-purple`. The third image is a red flower. So, it is named `red.jpg` and the folder containing the cut pieces of the image is named `cut-red`.

Now that we have taken care of the images, we can create our components.

First, we must remove `HelloWorld.vue` from the `src/components` folder since we don't need it anymore. We must also remove any reference to it from the `App.vue` file.

Next, in the `components` folder, we must create the `Puzzles.vue` file to let us select the puzzle. It has a template so that we can display the puzzles we select. In the `component options` object, we have an array that contains the puzzles data to display. Also, we have a method that lets us emit the event to our parent component, which is the `App.vue` component. This way, we can display the right puzzle in the slider puzzle component that we will create. To do that, in `src/components/Puzzles.vue`, we must add the following template code:

```
<template>
  <div>
    <h1>Select a Puzzle</h1>
    <div v-for="p of puzzles" :key="p.id" class="row">
      <div>
        <img :src="require(`../assets/${p.image}`)" />
      </div>
      <div>
        <h2>{{p.title}}</h2>
      </div>
      <div class="play-button">
        <button @click="selectPuzzle(p)">Play</button>
      </div>
    </div>
  </div>
</template>
```

Then, we must add the following script and style tags:

```
<script>
export default {
  data() {
    return {
      puzzles: [
        { id: 'cut-pink', image: "pink.jpg", title: "Pink
          Flower" },
        { id: 'cut-purple', image: "purple.jpg", title:
          "Purple Flower" },
        { id: 'cut-red', image: "red.jpg", title: "Red
          Flower" },
```

```
          ],
        };
      },
    ...
<style scoped>
.row {
  display: flex;
  max-width: 90vw;
  flex-wrap: wrap;
  justify-content: space-between;
}

.row img {
  width: 100px;
}

.row .play-button {
  padding-top: 25px;
}
</style>
```

In the `component options` object, we have the `data()` method, with the puzzle's reactive property between the script tags. It has an array of objects with the `id`, `image`, and `title` properties. The `id` property is a unique ID that we use when we render the entries with the `v-for` directive. We also emit the ID to `App.vue` so that we can pass it to our slide puzzle component from there as a prop. `title` is the title we display on the template in a human - readable way.

In the `methods` property, we have a `selectPuzzle()` method that takes the puzzle object. It calls `this.$emit` to emit the puzzle-changed event. The first argument is name. The second argument is the `payload` property that we want to emit in the event. We can listen to the event in the parent component by adding a `v-on` directive to the element wherever this component is referenced.

In the template, we have the `title` displayed with the `h1` component. The `v-for` directive loops through the items in the puzzle's `array` reactive property and displays them. As usual, we need the `key` prop for each entry to be set to a unique ID for Vue 3 to properly keep track of the values. We must also add a `class` attribute so that we can style the rows. To display the image, we can call `require` so that Vue 3 can resolve the path directly. The Vue CLI uses Webpack so that it can load the image as a module. We can set it as the value of the `src` prop and it will display the image. We load the whole images and display them.

Also, in the row, we have a button that calls the `selectPuzzle()` method when we click on it. This will set the choice of the puzzle and propagate it to the slider puzzle component that we will create so that we can see the correct puzzle displayed.

`.row img select` has its width set to `100px` to display a thumbnail of the whole image. Also, we can display the buttons in a way that they are aligned with the other child elements.

Next, we must create the `src/components/Records.vue` file to add a component that contains the speed records. This provides a list of fastest times for winning the game. The fastest time records are stored in local storage for easy access. In this component, all we do is display the components.

To create this component, we must write the following code in `src/components/Records.vue`:

```
<template>
  <div>
    <h1>Records</h1>
    <button @click="getRecords">Refresh</button>
    <div v-for="(r, index) of records" :key="index">{{
      index + 1}} - {{r.elapsedTime}}</div>
  </div>
</template>

<script>
export default {
  data() {
    return {
      records: [],
    };
  },
```

```
  created() {
    this.getRecords();
  },
  methods: {
    getRecords() {
      const records = JSON.parse(localStorage.getItem(
        "records")) || [];
      this.records = records;
    },
  },
};
</script>
```

In the `component` object, we have the `getRecords()` method, which obtains the fastest time records from local storage. The `localStorage.getItem()` method gets data by its key. The argument is the key that maps to the data that we want to get. It returns a string with the data. Therefore, to convert the string into an object, we must call `JSON.parse` to parse the JSON string into an object. It should be an array since we will create an array and stringify it into a JSON string before we record it. Local storage can only hold strings; so, this is a required step.

Once we've retrieved the records from local storage, we can set it as the value of the `this.records` reactive property. In case there is no item with the `records` key in local storage, we must set the default to an empty array. This way, we always get an array assigned to `this.records`.

Also, we have the `beforeMount` hook, which lets us get the records before the component mounts. This way, we will see the records when the component is mounted.

In the template, we show the speed records with the `v-for` directive to loop through the items and display them. The `v-for` directive in the array entry has the first item in the parentheses. The second item in the parentheses is the index. We can set the `key` prop to the index since they are unique, and we are not moving the entries around. We display both in the list.

Also, we have a button that calls the `getRecords` method when we click it to get the latest entries.

Now that we've created the simplest components, we can move on and create the slider puzzle component.

Creating the components for shuffling pictures

The slider puzzle game provides the slider puzzle where the player shuffles the tiles into a picture to win, the elapsed time display, the logic for rearranging the puzzles, the logic to check if we win, and a timer to calculate the elapsed time since the game started.

To calculate the elapsed time easily, we can use the moment library. To install the library, we can run `npm install moment`. Once we have installed the package, we can begin writing the necessary code.

Let's create the `src/components/SliderPuzzle.vue` file. The full code for this file can be found at `https://github.com/PacktPublishing/-Vue.js-3-By-Example/blob/master/Chapter03/src/components/SliderPuzzle.vue`.

We will start by creating the component with the `script` tag:

```
<script>
import moment from "moment";

const correctPuzzleArray = [
  "image_part_001.jpg",
  "image_part_002.jpg",
  "image_part_003.jpg",
  "image_part_004.jpg",
  "image_part_005.jpg",
  "image_part_006.jpg",
  "image_part_007.jpg",
  "image_part_008.jpg",
  "image_part_009.jpg",
];
...
</script>
```

First, we import the moment library to calculate the elapsed time. Next, we define the correctPuzzleArray variable and assign it to an array with the correct order of the files. We check against this array to determine if the player has won the game.

Then, we move on to creating the object for the component options. The `props` property contains our own prop. `puzzleId` is a string with the ID of the puzzle the player is playing. We must make sure that it is a string. We set its default value to `'cut-pink'` so that we always have a puzzle set.

The `data()` method contains our initial states. We return an object with them. This way, we make sure that the values of the reactive properties are always isolated from the other components in our app. The `correctPuzzleArray` reactive property is just what we defined earlier. We just set it to a property so that it becomes a reactive property. This makes it usable with our `isWinning` computed property since we want the value to update when this array updates:

```
<script>
...
export default {
  name: "SliderPuzzle",
  props: {
    puzzleId: {
      type: String,
      default: "cut-pink",
    },
  },
  data() {
    return {
      correctPuzzleArray,
      shuffledPuzzleArray: [...correctPuzzleArray].sort(
        () => Math.random() - 0.5
      ),
      indexesToSwap: [],
      timer: undefined,
      startDateTime: new Date(),
      currentDateTime: new Date(),
    };
  },
  ...
};
</script>
```

shuffledPuzzleArray is a copy of the correctPuzzleArray reactive property, but the items are shuffled so that the player has to rearrange the items to win the game. To create the value for the property, first, we must make a copy of the correctPuzzleArray array with the spread operator. Then, we must call sort with a callback. callback is a function that generates a number between -0.5 and 0.5 with Math.random() - 0.5. We need a random number between that range so that the values sort randomly. callback is a comparator function. It can take two parameters; that is, the previous and current array entry, so that we can compare them:

```
<script>
...
export default {
  ...
  computed: {
    isWinning() {
      for (let i = 0; i < correctPuzzleArray.length; i++) {
        if (correctPuzzleArray[i] !==
          this.shuffledPuzzleArray[i]) {
          return false;
        }
      }
      return true;
    },
    elapsedDiff() {
      const currentDateTime = moment(this.currentDateTime);
      const startDateTime = moment(this.startDateTime);
      return currentDateTime.diff(startDateTime);
    },
    elapsedTime() {
      return moment.utc(this.elapsedDiff).format(
        "HH:mm:ss");
    },
  },
};
</script>
```

Since we are sorting items randomly, we don't need to do any comparison. If the comparator callback returns a negative number or 0, then the order of the items is unchanged. Otherwise, the order of the items in the array we are sorting is switched around. The sort() method returns a new array with the entries sorted.

The indexesToSwap reactive property is used to add the index of the image filenames that we want to swap. When we click the swap() method, we push a new value to the indexesToSwap reactive property so that we can swap the two items with the given index when there are two items in the indexesToSwap array.

The timer reactive property may contain the object of the timer that's returned by the setInterval function. The setInterval function lets us run code periodically. It takes a callback with the code we want to run as the first argument. The second argument is the time between each call of the callback in milliseconds.

The startDateTime reactive property contains the date and time when the game started. It is a Date instance that contains the current time. The currentDateTime reactive property has the Date instance with the current date and time. It is updated as the game is processing within the callback property we pass into the setInterval function.

The data() method contains the initial values of all the reactive properties.

The computed property contains the computed properties. Computed properties are synchronous functions that return some values that are based on other reactive properties. Computed properties are reactive properties themselves. Their values are updated when the reactive properties that are referenced within the computed property functions that are referenced are updated. We defined three computed properties in this component: isWinning, elapsedDiff, and elapsedTime.

The isWinning computed property is the property that contains the state of the game. If it returns true, then the player wins the game. Otherwise, the player hasn't won the game. To check if the player has won the game, we loop through the correctPuzzleArray reactive property and check if each entry of it is the same as the one in the shuffledPuzzleArray reactive property array.

correctPuzzleArray contains the correct items listed. So, if each item in the shuffledPuzzleArray array's reactive property matches the entries in correctPuzzleArray, then we know that the player has won. Otherwise, the player hasn't won. Therefore, if there are any differences between correctPuzzleArray and shuffledPuzzleArray, then it returns false. Otherwise, it returns true.

The `elapsedDiff` computed property calculates the elapsed time in milliseconds. This is where we use the `moment` library to calculate the elapsed time from `startDateTime` to `currentDateTime`. We use the `moment` library to do this calculation since it makes our job a lot easier. It has a `diff()` method that we can use to calculate the difference between this and another `moment` object. The difference in milliseconds is returned.

Once we've calculated the `elapsedDiff` computed property, we can use it to format the elapsed with `moment` into a human-readable time format; that is, HH:mm:ss. The `elapsedTime` computed property has the computed property return a string with the formatted elapsed time. The `moment.utc()` method is a function that takes a timespan in UTC, then returns a `moment` object where we can call the `format()` method to let us calculate the time.

Now that we have defined all our reactive and computed properties, we can define our methods so that we can rearrange our slides into the correct picture.

Rearranging the slides

We can add the required `methods` for the `SliderPuzzle.vue` component by writing the following code:

```
<script>
...

export default {
  ...
  methods: {
    swap(index) {
      if (!this.timer) {
        return;
      }
      if (this.indexesToSwap.length < 2) {
        this.indexesToSwap.push(index);
      }
      if (this.indexesToSwap.length === 2) {
...
        this.resetTime();
        clearInterval(this.timer);
      },
```

```
    resetTime() {
       this.startDateTime = new Date();
       this.currentDateTime = new Date();
    },
    recordSpeedRecords() {
       let records = JSON.parse(localStorage.getItem(
          "records")) || [];
...
       localStorage.setItem("records", stringifiedRecords);
    },
  },
};
</script>
```

The logic is defined in the `methods` property. We have the `swap()` method to let us swap the cut image slides. The `start()` method lets us reset the reactive properties into their initial states, shuffle the cut photo slides, and then start the timer to calculate the elapsed time. We also check if the player has won each time the timer code is run. The `stop()` method lets us stop the timer. The `resetTime()` method lets us reset `startDateTime` and `currentDateTime` to their current date time. The `recordSpeedRecords()` method lets us record the time that the player took to win the game if they are in the top 10.

We start with the logic to swap the slides by defining the `swap()` method. It takes an argument, which is the index of one of the slides that we want to swap. When the player clicks on a slide, this method is called. This way, we add the index of one of the items we want to swap with the other to the `indexesToSwap` computed property. So, if the player clicks on two slides, then their positions will be swapped with each other.

The `swap()` method body checks if the `indexesToSwap` reactive property has less than two slide indexes inside it. If there's less than two, then we call `push` to append the slide to the `indexesToSwap` array. Next, if there are indexes in the `indexesToSwap` reactive property array, then we do the swapping.

To do the swapping, we destructure the indexes from the `indexToSwap` reactive property. Then, we use our destructuring assignment again to do the swapping:

```
[this.shuffledPuzzleArray[index1], this.
shuffledPuzzleArray[index2]] = [this.
shuffledPuzzleArray[index2], this.shuffledPuzzleArray[index1]];
```

To swap the items in an array, we just have to assign one with index2 of shuffledPuzzleArray to the item with index1. Then, the item that is originally in index1 of shuffledPuzzleArray is put into the index2 slot of shuffledPuzzleArray in the same way. Finally, we make sure that we empty the indexesToSwap array so that we can let the player swap another pair of slides. Since shuffledPuzzleArray is a reactive property, it is automatically rendered in the template as it updates with the v-for directive in the template.

The start() method lets us start the timer for calculating the elapsed time between when the **Start** button is clicked to start the game and the current date and time until the game is finished or when the user clicks the **Quit** button. First, the method resets the startDateTime and currentDateTime reactive properties by setting those values to the current date time, which we get by instantiating the Date constructor. Then, we shuffle the slides by making a copy of correctPuzzleArray, and then calling sort as we did previously to sort the copy of the correctPuzzle array. Also, we set the indexesToSwap property to an empty array to clear any items that are present so that we start afresh.

Once we've done all the resetting, we can call setInterval to start the timer. This will update the currentDateTime reactive property with the current date and time so that we can calculate the elapsedDiff and elapsedTime computed properties. Next, we check the isWinning reactive property to check if it is true. If it is, then we call the this.recordSpeedRecords method to record the fastest time if the player has won.

If the player wins, as indicated by isWinning being true, we can also call the stop() method to stop the timer. The stop() method just calls the resetTime() method to reset all the times. Then, it calls clearInterval to clear the timer.

To display the slider puzzle, we can add the template tag:

```
<template>
  <div>
    <h1>Swap the Images to Win</h1>
    <button @click="start" id="start-button">Start
      Game</button>
    <button @click="stop" id="quit-button">Quit</button>
    <p>Elapsed Time: {{ elapsedTime }}</p>
    <p v-if="isWinning">You win</p>
    <div class="row">
      <div
        class="column"
```

```
        v-for="(s, index) of shuffledPuzzleArray"
        :key="s"
        @click="swap(index)"
    >
        <img :src="require(`../assets/${puzzleId}/${s}`)"
        />
      </div>
    </div>
  </div>
</template>
```

Then, we can add the required styles by writing the following code:

```
<style scoped>
.row {
  display: flex;
  max-width: 90vw;
  flex-wrap: wrap;
}

.column {
  flex-grow: 1;
  width: 33%;
}

.column img {
  max-width: 100%;
}
</style>
```

In the `styles` tag, we have the styles for styling the slider puzzle. We need the slider puzzle so that we can display three slides in a row and three rows altogether. This way, we display all the slides in a 3x3 grid. The `row` class has the property set to `flex` so that we can use flexbox to lay out the slides. We also set the `flex-wrap` property to `wrap` so that we can wrap any overflowing items to the next row. `max-width` is set to `90vw` so that the slider puzzle grid will stay on the screen.

The `column` class has the `flex-grow` property set to `1` so that it is one of three items displayed in the row.

In the template, we display our `title` for the game with the `h1` element. We have a **Start Game** button that calls the `start()` method when we click on the button to start the game timer. Also, we have a **Quit** button to call the `stop()` method when we click on it to stop the timer. The `elapsedTime` computed property is displayed like any other reactive property. And if the user wins, as indicated by the `isWinning` reactive property returning true, we will see the **'You Win'** message.

To display the slides, we just loop through all the `shuffledPuzzleArray` reactive properties with the `v-for` directive and render all the slides. When we click on each slide, the `swap()` method is called with the index. And once we have two indexes in the `indexesToSwap` reactive property, we swap the slides. The `key` prop is set to the filename since they are unique. To display the slide images, we call `require` with the path of the image so that we display the images.

Since we have the flexbox styles to display the items three in a row and in three rows, all nine images will automatically be displayed in a 3x3 grid. Now that we have the slider puzzle game logic out of the way, all we have to add is the logic that records the timing score in local storage.

Calculating the score based on timing

This is done in the `recordSpeedRecords()` method. It gets the records by getting the local storage item with the *key* records from the local storage. Then, we get the `elapsedTime` and `elapsedDiff` reactive property values and push them into the `records` array.

Next, we sort the records with the `sort()` method. This time, we are not sorting the items randomly. Rather, we are sorting them by the `elapsedDiff` reactive property's timespan, which is measured in milliseconds. We pass in a callback with the a and b parameters, which are the previous and current array entries, respectively, and we return the difference between them. This way, if it returns a negative number or 0, then the order between them is unchanged. Otherwise, we switch the order. Then, we call `slice` with the first and last index to include it in the returned array that we assigned to the `sortedRecords` constant. The `slice()` method returns an array with the item in the first index included all the way up to the last index, minus 1.

Finally, we *stringify* the arrays with the `JSON.stringify()` method to convert the `sortedRecords` array into a string. Then, we call `localStorage.setItem` to put the item into an item with the `'records'` key.

Finally, we must change the contents of the App.vue file to the following:

```
<template>
  <div>
    <Puzzles @puzzle-changed="selectedPuzzleId = $event" />
    <Records />
    <SliderPuzzle :puzzleId="selectedPuzzleId" />
  </div>
</template>

<script>
import SliderPuzzle from "./components/SliderPuzzle.vue";
import Puzzles from "./components/Puzzles.vue";
import Records from "./components/Records.vue";

export default {
  name: "App",
  components: {
    SliderPuzzle,
    Puzzles,
    Records,
  },
  data() {
    return {
      selectedPuzzleId: "cut-pink",
    };
  },
};
</script>
```

We add the components we created earlier to render them on the screen. selectedPuzzleId has the ID of the puzzle we selected by default.

Now that we have all the code, we can run the project by running npm run serve in our project folder if we haven't already. Then, when we go to the URL that is indicated by the Vue CLI, we will see the following:

Figure 3.3 – Screenshot of the slider puzzle game

Now that we've finished the web app's code, we have to find an easy way to test all its parts.

Unit testing with Jest

Testing is an important part of any app. When we refer to tests, we usually mean automated tests. These are tests that we can run in quick repetition to make sure that our code is not broken. When any tests fail, we know that our code did not do what it was doing before. Either we created a bug, or the tests are outdated. Because we can run them quickly, we can write many of them and run them as we build our code.

This is much preferred to manual tests, which must be done by a person doing the same actions over and over again. Manual tests are boring for the tester, they are error-prone, and are very slow. It is just not a pleasant experience for anyone. Therefore, it is better to write as many automated tests as possible to minimize the manual tests.

If the instructions that are shown in the Vue CLI are followed, it is very easy to add skeleton test code without doing any extra work. The files for unit tests should be automatically generated for us. We should have a `tests/unit` folder in our code to separate our test code from our production code.

Jest is a JavaScript test framework that we can run unit tests with. It provides us with a useful API that lets us describe our tests groups and define our tests. Also, we can mock any external dependencies that are normally used, such as timers, local storage, and states, easily. To mock the `localStorage` dependency, we can use the `jest-localstorage-mock` package. We can install it by running `npm install jest-localstorage-mock --save-dev`. The `--save-dev` flag lets us save the package as a development dependency so that it is only installed in the development environment and nowhere else. Also, in the `package.json` file, we will add a `jest` property to it as a `root` property. To do that, we can write the following code:

```
"jest": {
"setupFiles": [
"jest-localstorage-mock"
  ]
}
```

We have these properties in `package.json` so that when we run our tests, the `localStorage` dependency will be mocked out so that we can check if its methods have been called. Together with the other properties, our `package.json` file should look something like the following:

```
{
  "name": "vue-example-ch3-slider-puzzle",
  "version": "0.1.0",
  "private": true,
  "scripts": {
    "serve": "vue-cli-service serve",
    "build": "vue-cli-service build",
    "test:unit": "vue-cli-service test:unit",
    "lint": "vue-cli-service lint"
  },
```

```
    "dependencies": {
      "core-js": "^3.6.5",
      "lodash": "^4.17.20",
      "moment": "^2.28.0",
      "vue": "^3.0.0-0"
    },
    "devDependencies": {
...
      "eslint-plugin-vue": "^7.0.0-0",
      "jest-localstorage-mock": "^2.4.3",
      "typescript": "~3.9.3",
      "vue-jest": "^5.0.0-0"
    },
    "jest": {
      "setupFiles": [
        "jest-localstorage-mock"
      ]
    }
}
```

Once we are done with that, we can add our tests.

Adding a test for the Puzzles.vue component

First, we must remove the existing files from the tests/unit folder. Then, we can start writing our tests. We can start by writing tests for the Puzzles.vue component. To do that, we must create the tests/unit/puzzles.spec.js file and write the following code:

```
import { mount } from '@vue/test-utils'
import Puzzles from '@/components/Puzzles.vue'

describe('Puzzles.vue', () => {
  it('emit puzzled-changed event when Play button is
    clicked', () => {
    const wrapper = mount(Puzzles)
    wrapper.find('.play-button button').trigger('click');
    expect(wrapper.emitted()).toHaveProperty('puzzle-
```

```
        changed');
    })

    it('emit puzzled-changed event with the puzzle ID when
        Play button is clicked', () => {
        const wrapper = mount(Puzzles)
        wrapper.find('.play-button button').trigger('click');
        const puzzleChanged = wrapper.emitted('puzzle-
            changed');
        expect(puzzleChanged[0]).toEqual([wrapper.vm.puzzles[0].id]
    );
    })
})
```

The `describe` function takes a string with the description of our test group in a string. The second argument is a callback with the tests inside it. The `describe` function creates a block that groups several related tests together. Its main purpose is to make the test results easier to read on our screens.

The `it()` function lets us describe our tests. It is also known as the `test()` method. Its first argument is the `name` property of the test in string form. The second argument is a callback function with the test code. It also takes an optional third argument with `timeout` in milliseconds so that our tests won't be stuck running forever. The default timeout is 5 seconds.

If a `promise` is returned from the `it` or `test` function, Jest will wait for the promise to resolve before the test completes. Jest also waits if we provide an argument to the `it` or `test` function, which is usually called `done`. The `done` function is called to indicate that the test is done if the `done` parameter is added to the `it` or `test` callback.

The `it` or `test` function doesn't have to be inside the callback we pass into `describe`. It can also be called **standalone**. However, it is better to group related tests together with `describe` so that we can read the results more easily.

The first test tests that when the **Play** button is clicked, the `puzzle-changed` event is emitted. As we can see from the `Puzzles.vue` component, the `puzzle-changed` event is emitted with the `this.$emit()` method. To create our test, we call `mount` to mount our component. It takes the component we want to test as its argument. It also takes a second argument with the object of component options that we want to override. In this test, since we are not overriding anything, we did not pass in anything as the second argument.

The mount () method returns the wrapper object, which is the wrapper object for our component that we are testing. It has a few handy methods that we can use to do the testing. In this test, we call the find () method to get the HTML element with the given selector. It returns the HTML DOM object, which will call the trigger () method to trigger the event that we want in our test.

This way, we can trigger events such as keyboard and mouse events to simulate user interaction. So, the following code is used to get the element with the .play-button button selector and then trigger the click event on it:

```
wrapper.find('.play-button button').trigger('click');
```

The last line of the test is used to check whether the puzzle-changed event is emitted. The emitted () method returns an object with properties that have names. These are the event names of the emitted events. The toHaveProperty () method lets us check if the property name we passed in as the argument is in the returned object. It is a property of the object that's returned by the expect () method.

In the second test, we mount the component again and trigger the click event on the same element. Then, we call the emitted () method with the event name so that we can get the payload that is emitted with the event with the object it returns. The puzzleChanged array contains the payload that is emitted as the first element. Then, to check if the puzzles [0] . id property is emitted, we have the check in our last line. The wrapper . vm property contains the mounted component object. Therefore, wrapper . vm.puzzles is the puzzle's reactive property of the Puzzles component. So, this means we are checking if the id property of the puzzle's reactive property from the Puzzles component has been emitted.

Adding a test for the Records component

Next, we must write tests for the Records component. To do that, we must create the tests/unit/records.spec.js file and write the following code:

```
import { shallowMount } from '@vue/test-utils'
import 'jest-localstorage-mock';
import Records from '@/components/Records.vue'


describe('Records.vue', () => {
  it('gets records from local storage', () => {
    shallowMount(Records, {})
    expect(localStorage.getItem).
      toHaveBeenCalledWith('records');
```

```
    })
  })
```

This is where we use the `jest-localstorage-mock` package. All we have to do is to import the package file; then, the code in the file will run and mock the `localStorage` dependency for us. In the test, we call `shallowMount` to mount our `Records` component and then we can check if `localStorage.getItem` is called with the `'records'` argument. With the `jest-localstorage-mocks` package, we can pass in `localStorage.getItem` directly to expect it to do the check. The `toHaveBeenCalledWith()` method lets us check the argument that it is called with.

Since we called the `localStorage.getItem()` method in the `beforeMount()` method, this test should pass since we called it as we were loading the component.

Adding a test for the SliderPuzzle component

Finally, we must write some tests for the `SliderPuzzle` component. We will add the `tests/unit/sliderPuzzle.spec.js` file and write the following code:

```js
import { mount } from '@vue/test-utils'
import SliderPuzzle from '@/components/SliderPuzzle.vue'
import 'jest-localstorage-mock';
jest.useFakeTimers();

describe('SliderPuzzle.vue', () => {
  it('inserts the index of the image to swap when we click
    on an image', () => {
    const wrapper = mount(SliderPuzzle)
    wrapper.find('#start-button').trigger('click')
...
    expect(firstImage).toBe(newSecondImage);
    expect(secondImage).toBe(newFirstImage);
  })


  ...
  })


  afterEach(() => {
    jest.clearAllMocks();
```

```
    });
})
```

In the `'inserts the index of the image to swap when we click on an image'` test, we mount the `SliderPuzzle` component and then trigger the `click` event on the `img` element. The `img` element is the first slide of the slider puzzle. The `swap()` method should be called so that the `indexesToSwap` reactive property has the index of the first image that's added. The `toBeGreaterThan()` method lets us check if the returned value of what we expected is greater than some number.

In the `'swaps the image order when 2 images are clicked'` test, we mount the `SliderPuzzle` component again. Then, we get `wrapper.vm.shuffledPuzzleArray` to get the indexes that are in the earlier array and destructure their values. We will use it later to compare the values from the same array to see if they have been swapped once we've clicked on two images.

Next, we trigger the click on the slides with the `wrapper.get()` method to get the image element. Then, we call the `trigger()` method to trigger the click events. Then, we check if the `indexesToSwap` reactive property has 0 for its length after the swapping is done. Then, in the last three lines, we get the items from `wrapper.vm.shuffledPuzzleArray` again and compare their values. Since the entries are supposed to be swapped after the two slides, we have the following code to check if swapping is actually done:

```
expect(firstImage).toBe(newSecondImage);
expect(secondImage).toBe(newFirstImage);
```

In the `'starts timer when start method is called'` test, we mount the `SliderPuzzle` component again. This time, we call the `start()` method to make sure that the timer is actually created with `setInterval`. We also check if the `setInterval` function is called with a function and 1,000 milliseconds. To let us test anything with timers easily, which includes testing anything that calls `setTimeout` or `setInterval`, we call `jest.useFakeTimers()` to let us mock those functions so that our tests won't interfere with the operations of the other tests:

```
import { mount } from '@vue/test-utils'
import SliderPuzzle from '@/components/SliderPuzzle.vue'
import 'jest-localstorage-mock';
jest.useFakeTimers();


describe('SliderPuzzle.vue', () => {
```

```
...

  it('starts timer when start method is called', () => {
    const wrapper = mount(SliderPuzzle);
    wrapper.vm.start();
    expect(setInterval).toHaveBeenCalledTimes(1);
    expect(setInterval).toHaveBeenLastCalledWith(expect.any(
      Function), 1000);
  })

...

  afterEach(() => {
    jest.clearAllMocks();
  });
})
```

The `toHaveBeenCalledTimes()` method checks whether a function that we passed into the `expect()` method is called a given number of times. Since we called `jest.useFakeTimers()`, `setInterval` is actually a spy of the real `setInterval` function rather than the real version. We can only use spies for a function with `expect` and `toHaveBeenCalledTimes` and `toHaveBeenCalledWith`. So, the code we have will work. The `toHaveBeenLastCalledWith()` method is used to check the argument that our function spy is called with the given kind of argument. We make sure the first argument is a function and that the second argument is 1,000 milliseconds.

In the `'stops timer when stop method is called'` test, we do something similar by mounting the component and then calling the `stop()` method. We make sure `clearInterval` is actually called when we call the `stop()` method:

```
import { mount } from '@vue/test-utils'
import SliderPuzzle from '@/components/SliderPuzzle.vue'
import 'jest-localstorage-mock';
jest.useFakeTimers();

describe('SliderPuzzle.vue', () => {
  ...

  it('stops timer when stop method is called', () => {
    const wrapper = mount(SliderPuzzle);
    wrapper.vm.stop();
```

```
      expect(clearInterval).toHaveBeenCalledTimes(1);
  })

  it('shows the elapsed time', () => {
    const wrapper = mount(SliderPuzzle, {
      data() {
        return {
          currentDateTime: new Date(2020, 0, 1, 0, 0, 1),
          startDateTime: new Date(2020, 0, 1, 0, 0, 0),
        }
      }
    });
    expect(wrapper.html()).toContain('00:00:01')
  })

  ...

  afterEach(() => {
    jest.clearAllMocks();
  });
})
```

Next, we add the `'records record to local storage'` test. We make use of the
`jest-localstorage-mock` library again to mock the `localStorage` dependency.
We mount the `SliderPuzzle` component differently in this test. The second argument
is an object that contains the `data()` method. This is the `data()` method that we have
in the component's `options` object. We override the component's original reactive
property values with what we pass in. The `currentDateTime` and `startDateTime`
reactive properties are overridden so that we can set the date to what we want so that we
can do the testing with them instead.

Then, we call the `wrapper.vm.recordSpeedRecords()` method to test if the
`localStorage.setItem()` method is called. We call the method that is in the
mounted component. Then, we create the `stringifiedRecords` JSON string so
that we can compare that with what is being called with `localStrorage.setItem`.
`toHaveBeenCalledWith` only works with `localStorage.setItem` because we
imported the `jest-localstorage-mock` library to create a spy from the actual
`localStorage.setItem()` method. This lets Jest check whether the method is called
or not with the given arguments.

To test if the timer is started when the **Start** button is clicked, we have the `'starts timer with Start button is clicked'` test. We just get the **Start** button by its ID with the `get()` method and trigger the `click` event on it. Then, we check that the `setInterval` function is called. Like with `localStorage`, we mock the `setInterval` function with the `jest.useFakeTimers()` method to create a spy from the actual `setInterval` function. This lets us check that it is called.

Similarly, we have the `'stops timer with Quit button is clicked'` test to check if the `clearInterval` function is called if the **Quit** button is clicked.

Finally, we have the `'shows the elapsed time'` test to mount the component with different values for the `currentDateTime` and `startDateTime` reactive properties. They are set to the values we want, and they will stay the way they are in the test. Then, to check if the `elapsedTime` computed property is displayed properly, we call the `wrapper.html()` method to return the rendered HTML in the wrapped component, and we check that it includes the elapsed time string we are looking for.

To clean up the mocks after each test so that we start afresh after each test, we call the `jest.clearAllMocks()` method to clear all the mocks after each test. The `afterEach` function takes a callback that is run after each test is done.

Running all the tests

To run the tests, we run `npm run test:unit`. By doing this, we'll see something like the following:

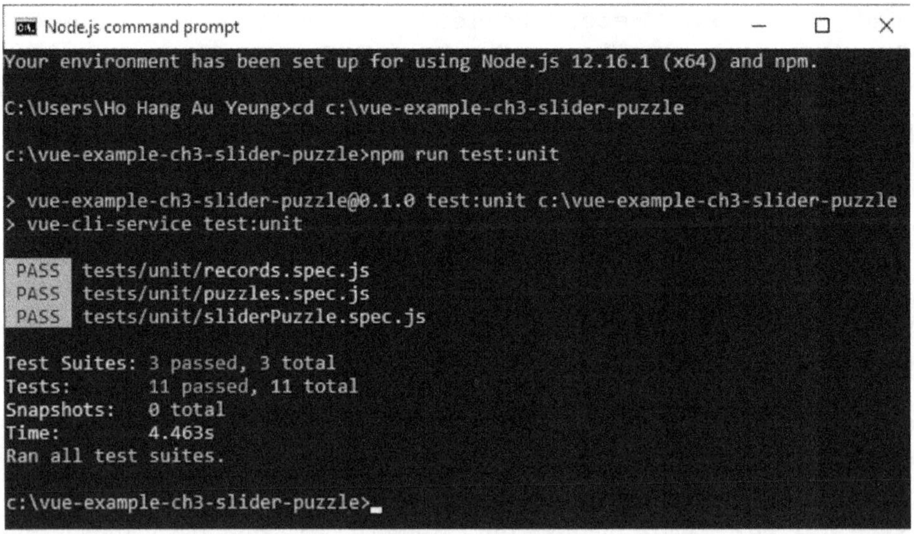

Figure 3.4 – Results of our unit tests

Since all the tests passed, the code in our project is doing what we expect it to. It only takes around 4 seconds to run all the tests, which is much faster than testing our code manually.

Summary

In this chapter, we looked deeper into components by defining the computed properties in our components. Also, we added tests for our components so that we can test the parts of our components individually. With the Vue CLI, we added test files and dependencies easily within our app.

Inside our components, we can emit events that propagate to the parent component with the `this.$emit()` method. It took a string with the event name. The other arguments are the payloads that we want to pass from the parent component to the child components.

To add unit tests to our Vue 3 app and run the tests, we used the Jest test framework. Vue 3 adds its own specific APIs to Jest so that we can test Vue 3 components with it. To test components, we mounted the component by using the `mount` and `shallowMount` functions. The `mount` function lets us mount the component itself, including the nested component. The `shallowMount` function only mounts the component itself without the child components. They both return a `wrapper` for our component so that we can use it to interact with the component to do the testing.

We should make sure that our tests run in isolation. This is why we were mocking the external dependencies. We do not want to run any code that requires anything external to the tests and the project code to be available. Also, we had to make sure that we cleaned up any dependencies in our tests if needed. If there were any mocks, we had to clean them up so that they were not carried forward to another test. Otherwise, we may have tests that depend on other tests, which makes troubleshooting tests very difficult.

In the next chapter, we will look at how to create a photo gallery app that saves data by sending the data we want to save to a backend API. We will introduce the use of Vue Router so that we can navigate to different pages.

4
Building a Photo Management Desktop App

So far, we have only built web applications with Vue 3. In this chapter, we will build a photo management desktop app with the Vue Electron plugin. We will learn how to easily build cross-platform desktop apps with Electron and Vue. This is useful because we can build cross-platform desktop apps without much additional effort. This will save us time and get us good results.

In this chapter, we will focus on the following topics:

- Understanding components
- Creating a project with Vue CLI Plugin Electron Builder
- Adding a photo submission UI
- Adding a photo display
- Adding routing to the Photo Manager app
- Using our app with photo management APIs

Technical requirements

To understand this chapter, you should already know how to do the following:

- Create basic Vue components
- Create projects with the Vue CLI

You can find all the code for this chapter at `https://github.com/ PacktPublishing/-Vue.js-3-By-Example/tree/master/Chapter04`.

Understanding components

Components can only have so much inside them. They take props from the parent component, so we can customize their behavior. Additionally, they can have computed properties and watchers to watch for reactive properties and return the data or do the things we want them to do. They can also have methods that allow us to do specific things with them. Components should be simple; that is, they should not have too much going on inside them. Templates should only have a few elements and components inside them in order to keep them simple. Components also have some built-in directives for us to manipulate the **Document Object Model** (**DOM**) and carry out data binding.

Other than that, components cannot do much. It would be impossible to build anything with non-trivial complexity if we only have a few components and no way to navigate with URLs.

If our app only has components, then we can only nest a few of them before it gets too complex. Additionally, if we have lots of nested components, then the navigation becomes difficult. Most apps have different pages, and it's impossible to navigate without some kind of routing mechanism in place.

With Vue Router, we can render the components we want when we go to a given URL. We can also navigate to the routes with the `router-link` component provided by Vue Router. Vue Router has many features. It can match URLs to routes. The URL can have query strings and URL parameters. Additionally, we can add nested routes with it so that we can nest the routes inside different components. Components that are mapped to URLs are displayed in Vue Router's `router-view` component. If we have nested routes, then we need the `router-view` component in the parent route. This way, the child routes will be displayed.

To navigate to different routes, Vue Router provides a wrapper for the JavaScript History API, which is built into nearly all modern browsers. With this API, we can easily go back to a different page, go to a historical record, or go to the URL we want. Vue Router also supports HTML5 mode so that we can have URLs that don't have the hash sign to distinguish them from server-side rendered routes.

Additionally, transition effects are supported, and we can see them when we navigate between different routes. Styles can also be applied to the links when the link is the one that has been navigated to and is active.

Understanding Vue CLI Plugin Electron Builder

We can convert JavaScript client-side web applications into desktop apps with Electron. Vue CLI Plugin Electron Builder enables us to add the files and settings to build a Vue 3 app inside a desktop app without many manual changes. Essentially, an Electron app is a web app that runs inside a Chromium browser wrapper that displays our web app. Therefore, it can do anything that we need it to do, with a browser. This includes some limited hardware interaction such as using microphones and cameras. Also, it provides some native capabilities such as displaying items in the notification area of popular operations systems and displaying native notifications. It is intended to be an easy way to build desktop apps from web apps that do not require low-level hardware access.

Vue CLI Plugin Electron Builder is the fastest way to create an Electron app from a Vue app, as it has support for some native code. We can also include native modules in our code with it; we just have to include the locations of the Node.js modules with the native code that we want to include. Additionally, we can include environment variables to build our code accordingly for different environments. Web workers are also supported since this is supported by Chromium. We can use it to run background tasks so that we don't have to hold up the main browser thread when we want to do long-running or CPU-intensive tasks. All this can be done with either regular Vue app code or via configuration changes. There are also other things that Electron supports that are not supported with Vue CLI Plugin Electron Builder. This means it's more limited in its capabilities. Features such as native menus are not available when using this plugin. However, we can build the desktop app on multiple platforms.

In this chapter, we will build a photo manager desktop app that will run on Windows. The app will consist of a page to display all the photos you have added. Also, it will allow users to add photos and store them; it will have a form to enable users to add photos. The photos will be stored using our own photo storage mechanism. We will use Vue Router to let us navigate through the pages manually or automatically.

Creating a project with Vue CLI Plugin Electron Builder

Creating a project with Vue 3 and Vue CLI Plugin Builder is an easy task. Perform the following steps:

1. To create the project, create a project folder called `vue-example-ch4-photo-manager`. Then, go into the folder and run the following commands with npm:

   ```
   npm install -g @vue/cli@next
   npm vue create
   ```

   Alternatively, you can run the following commands with `yarn`:

   ```
   yarn global add @vue/cli@next
   yarn create
   ```

   Once you have run the command, you should see a menu with choices for the types of projects you can create.

2. Pick the default **Vue 3 project** option to create a Vue 3 project. Then, add Vue CLI Plugin Electron Builder to our Vue app.

3. To add the Vue CLI Electron Builder plugin, run `vue add electron-builder` in your project folder. Run `cd <folder path>` to navigate to the folder. All the files and settings will be added to the project.

Once the command is run, we should see a couple of new things in our project. We now have the `background.js` file that has the code to display the Electron window. We also have a few new script commands added to our `package.json` file. The `electron:build` script command lets us build our app for production. The `electron:server` command lets us serve our Electron app with a development server so that we can develop our app with it. We are automatically provided with hot reload capability so that we can see all the changes live in the browser and also in the Electron desktop app window. Since it's just a wrapper for a browser, we can see the latest changes in the desktop window with our app.

In the desktop window, we should also see the Chromium development console, which is the same as the one in the browser. To make debugging easier, we suggest looking at the browser window for debugging since we can inspect elements and do whatever we want to do in our development console.

The contents of the `background.js` file can be found at `https://github.com/PacktPublishing/-Vue.js-3-By-Example/blob/master/Chapter04/src/background.js`.

The `BrowserWindow` constructor creates a browser window with a width of 800 px and a height of 600 px, by default. We can change the window size by dragging the window as we would do with other desktop apps. The `win.loadURL()` method loads the home page for our app, which is the `index.html` file. The `index.html` file is in the `dist-electron` folder. Otherwise, we call the `win.loadURL()` method to load the `webpack-dev-server` URL in order to see the app in a window in the development environment. This is because `webpack-dev-server` only loads the app in memory when it is run in the development environment.

The `win.webContents.openDevTools()` method opens the Chromium development console in the development environment. The `app.quit()` method exits the desktop app. We listen to the message event when it is running under Windows, as indicated by the `process.platform === 'win32'` expression. Otherwise, Electron listens to the SIGTERM event and closes the window when that is emitted. The SIGTERM event is emitted when we end a program.

To change the title of the window, we just change the `title` tag in the `public/index.html` file. For example, we can write the following:

```
<!DOCTYPE html>
<html lang="en">
  <head>
    <meta charset="utf-8">
    <meta http-equiv="X-UA-Compatible" content="IE=edge">
    <meta name="viewport" content="width=device-width,
        initial-scale=1.0">
    <link rel="icon" href="<%= BASE_URL %>favicon.ico">
    <title>Photo App</title>
  </head>
  <body>
    <noscript>
      <strong>We're sorry but <%= htmlWebpackPlugin.options
        .title %> doesn't work properly without JavaScript
          enabled. Please enable it to continue.</strong>
    </noscript>
    <div id="app"></div>
    <!-- built files will be auto injected -->
  </body>
</html>
```

We just change the `title` tag's content to what we want, and the text will be displayed as the title of the window.

Then, once we run the `vue add electron-builder` command, we get the files and settings added. To start the development server and display the Electron app on our screen, we run the `yarn electron:serve` or `npm run electron:serve` commands, which are provided by Vue CLI Plugin Electron Builder. You should see a window display on the screen (please refer to *Figure 4.1*). This will automatically refresh when we make any changes to the existing files or if we add or remove files. Now we are almost ready to start building our app:

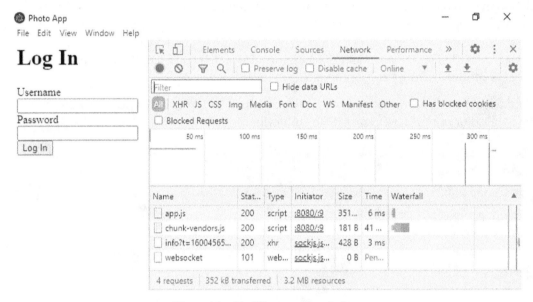

Figure 4.1 – The Electron app window

We will install a few packages that we need to use in our Vue app and the Vue Router library to add routing to our app. We will use it for the router links and to also navigate programmatically. We will also use the Axios HTTP client to make HTTP requests easily to our API. To install the packages, we can run the `npm install axios vue-router@4.0.0-beta.9` or `yarn add axios vue-router@4.0.0-beta.9` commands.

We install Axios so that we can make HTTP requests conveniently. Now we are ready to build our app.

Adding a photo submission UI

To build the app, we will first add our components and the file to store the constants that we will need throughout the app. To start with, we get rid of the components/ HelloWorld.vue file. Then, we remove any references of it in the App.vue file, including the import and components properties to unregister the component. Now we will add some new components to our app.

First, we add the PhotoFormPage.vue component file to the src/components folder. Inside the file, we add the following template:

```
<template>
  <div class="form">
    <h1>{{ $route.params.id ? "Edit" : "Add" }} Photo</h1>
    <form @submit.prevent="submit">
      <div>
        <label for="name">Name</label>
        <br />
        <input
          type="text"
          v-model="form.name"
          name="name"
          id="name"
          class="form-field"
        />
...
        <label for="dateTaken">Date Taken</label>
        <br />
        <input
          type="datetime-local"
          name="dateTaken"
          id="dateTaken"
          v-model="form.dateTaken"
        />
      </div>
      ...
    </form>
  </div>
</template>
```

This template has the inputs for adding and editing photos. `name` and `description` are text inputs. The `Date Taken` field is a date input.

Then, we add the photo field as a file input:

```
<template>
  <div class="form">
    <h1>{{ $route.params.id ? "Edit" : "Add" }} Photo</h1>
    <form @submit.prevent="submit">
      ...
      <div>
        <label for="photoFile">Photo</label>
        <br />
        <input type="file" name="photoFile" id="photoFile"
          @change="onChange" />
        <br />
        <img :src="form.photoFile" id="photo-preview" />
      </div>
    </form>
  </div>
</template>
```

We will read the selected file into a base64 string so that we can save the data easily using an HTTP request. We also use the base64 string to preview the image in the `img` element.

Then, we add the `script` tag to `PhotoFormPage.vue`:

```
<script>
import axios from "axios";
import { APIURL } from "../constant";

export default {
  name: "PhotoForm",
  data() {
    return {
      form: {
        name: "",
```

```
          description: "",
          dateTaken: "",
          photoFile: undefined,
        },
      };
    },
    ...
      reader.onload = () => {
        this.form.photoFile = reader.result;
      };
    },
  },
  async beforeMount() {
    const { id } = this.$route.params;
    if (id) {
      const { data } = await
        axios.get(`${APIURL}/photos/${id}`);
      this.form = data;
    }
  },
};
</script>
```

This determines whether we are editing an existing photo or creating a new one by checking whether the id parameter is set. If it is, then we are editing. Otherwise, we are creating a new photo. This logic is used in the submit method and the beforeMount hook.

The submit () method's id is used to make an HTTP request to the API to save the entry. We get the id parameter from the route by writing the following:

```
const { id } = this.$route.params;
```

Then, we add an if statement immediately below that to check whether it is set. If it is set, we make a *PUT* request to update an existing entry. Otherwise, we make a *POST* request to create a new entry.

In this component, we have a form that allows us to add and edit the photos in our app. We display the **Add** or **Edit** text depending on the value of the `edit` prop. Then, we have a form with a bunch of fields in it. The `form` element has the submit event listener that runs the `submit()` method when we click on the **Submit input** button. The `prevent` modifier runs the `event.preventDefault()` method without us having to add it ourselves inside the `submit` handler. We need this since we don't want the browser to execute the default `submit` behavior, which will submit our form data to the server directly. We want to process the data on the client side with our own client-side code.

In this project, we won't create our own API and we won't do any data validation on the server side.

Also, the modifier saves us from typing it in and also makes our code shorter. The directive syntax is common enough that there is a symbol for it. The `@` symbol can also be replaced with the `v-on` directive since `@` is the shorthand for the `v-on` directive.

Inside the `form` tag, we have the `input` elements with the `v-model` directive that binds to various properties of the `form` reactive property. The `label` HTML element is the label of each input. The `label` has the `for` attribute, which allows it to map the label to the `id` parameter of the `input` element. This is good for accessibility since screen readers will pick it up and read it to the user. This is very helpful for visually impaired users of our app. We will use very similar code in the `textarea` tag.

The date and time picker is a native date and time picker that is created by setting the `type` attribute to `datetime-local`. This enables us to add a date and time picker that is set to the time zone of your device. Then, we set the `v-model` directive to bind the date and time picker value to the one that the user picked in the browser or desktop app window. Most modern browsers support this type of input, so we can use this to enable users to pick the date and time. The `type` attribute can also be set to `date` to add a date picker only. Additionally, we can set the type to `datetime` to add a date and time picker that is set to UTC.

The `file` input is more complex. The input's `type` attribute is set to `file` so that we can see a file input. Additionally, it has a change event listener that runs the `onChange()` method to convert the binary image file into a base64 string. This saves the image to our API as a text string. For a small app, such as this photo management app, we can save images directly as a string.

However, if we were building a production-quality app with lots of users using the app and making lots of file uploads, then it would be a good idea to save the files in a third-party storage service, such as Dropbox or Amazon S3. Then, we could just get the files from the URL instead of as a base64 string. The HTTP URL and the base64 URL are equivalent. We can set both as the value of the `src` attribute of the `img` tag to display the image. In this case, we set the `src` attribute to the base64 URL in the `img` tag inside our form element.

At the bottom of our form, we have the input with the `type` attribute set to `submit`. This allows us to submit the input by pressing *Enter* on a form input element or clicking on the **Submit input** button.

Next, we add the `data()` method. This returns the initial values of the `form` reactive property. The `form` reactive property includes the `name`, `description`, `dateTaken`, and `photoFile` properties. The `name` property is the name of our photo. The `description` property is the description of our photo. The `dateTaken` property has a string with the date and time that the photo was taken at. And the `photoFile` property is the base64 string representation of the photo file.

Next, we have a few methods in this component. First, we have the `submit()` method that either makes a *PUT* request, to update an existing photo entry, or a *POST* request, to create a new photo entry. Before we make any HTTP requests, we check whether all the properties of the `this.form` reactive property are populated with non-falsy values. We want all the fields to be filled in. If there is a false value that has been set as the value of any of the properties, then we show an alert that tells the user to fill in all the fields.

To make the process of getting the properties shorter, we destructure the properties of the `this.form` reactive property and then carry out a check. After that, we check whether the `edit` prop is `true`. If it is, then we use the *PUT* request to update an existing entry. The `id` prop is set to the `$route.params.id` value so that we get the value of the ID URL parameter from the URL.

If the `edit` reactive property is `true`, then we make a *PUT* request to our API to update an existing photo entry. To make the *PUT* request, we call the `axios.put()` method. This takes the URL as the first argument and an object with the request body content as the second argument. Otherwise, we call `axios.post()` with the same arguments to make a *POST* request to create a new photo entry. The URL of the *PUT* request has the ID of the photo entry attached to the end of it so that the API can identify which entry to update:

```
{
  methods: {
    async submit() {
      const { name, description, dateTaken, photoFile } =
        this.form;
      if (!name || !description || !dateTaken ||
        !photoFile) {
        alert("All fields are required");
        return;
      }
```

```
    if (this.edit) {
      await axios.put(`${APIURL}/photos/${this.id}`,
        this.form);
    } else {
      await axios.post(`${APIURL}/photos`, this.form);
    }
    this.$router.push("/");
  },
  onChange(ev) {
    const reader = new FileReader();
    reader.readAsDataURL(ev.target.files[0]);
    reader.onload = () => {
      this.form.photoFile = reader.result;
    };
  },
  }
}
```

We have also defined the onChange() method to be used as the change event listener for the file input. When we select a file, this method is run. In the method body, we create a new FileReader instance to read the selected image file into a base64 string. The parameter has an event object, which has the file that we selected. The ev.target. files property is an array-like object with the selected files. Since we only allow the user to select one file, we can use the 0 property to get the first file. 0 is a property name and not an index since the files property is an array-like object; that is, it just looks like an array but doesn't act like it. However, it is an iterative object, so we can use the for-of loop or the spread operator to loop through the items or convert them into an array.

To read the selected file into a base64 string, we call the reader.readAsDataURL method with the file object as the argument to read the file into a base64 string. Then, we get the result by listening to the load event emitted by the reader object. We do this by setting an event handler as the value of the onload property. The result that is read is inside the result property. It is set to the this.form.photoFile property so that we can display the image in the img tag below our file input and also store it in our database after we submit it.

Then, we add some code to the `beforeMount` hook. We check for the value of the `this.edit` prop and then get the photo entry from our API if the value of the `this.edit` prop is `true`. We only need to check when we are mounting this component since we are using this component in a `route` component. Additionally, a `route` component is mounted when we go to a URL that maps to the component. When we go to another URL, the component will be unmounted. Therefore, we won't need a watcher to watch for the values of the `edit` or `id` props. We set the retrieved data to the `form` reactive property so that the user can see the data in the form fields and edit them as they wish.

The `axios.post()`, `axios.put()`, and `axios.get()` methods all return a promise that resolves to the response data as its resolved value. The `data` property has the response data. Therefore, we can use the `async` or `await` syntax to make our promise code shorter, just as we did in the entire component.

In the `style` tag, we have several styles that we can use to style our form. We display the form closer to the center of the screen by adding a `margin` property and setting it to `0 auto` in our `form` class. The width is set to `70vw` so that it is set to take up only 70 percent of the viewport width instead of the entire width. The `form-field` class has the `width` property set to 100 percent so that the form fields fill up the entire width of the form. Otherwise, they will be displayed at the default width, which is very short. The `photo-preview` ID is assigned to the `img` tag that we use for the preview. And we set the `width` property of it to `200px` so that we only show a thumbnail preview of the image:

```
<style scoped>
.form {
    margin: 0 auto;
    width: 70vw;
}

.form-field {
    width: 100%;
}

#photo-preview {
    width: 200px;
}
</style>
```

In this file, we make requests that allow us to edit or delete the photo entry.

Next, we create a component for our home page. We will create a `HomePage.vue` file in the `src/components` folder and write the following code in it:

```
<template>
  <div>
    <h1>Photos</h1>
    <button @click="load">Refresh</button>
    <div class="row">
      <div>Name</div>
      <div>Photo</div>
      <div>Description</div>
      <div>Actions</div>
    </div>
...
<script>
import axios from "axios";
import { APIURL } from "../constant";

export default {
  data() {
    return {
      photos: [],
    };
  },
  methods: {
    ...
  }
</script>

<style scoped>
  ...
</style>
```

This file is more complex than the components we previously created. In the `component options` object, we have the `data()` method that returns the initial values for our reactive properties. We only have one in this component to hold the photos. The `photos` reactive property will have the files. In the `methods` property, we have a few methods that can be used to populate the reactive properties. The `load` method uses the `axios.get()` method to get the data from the `photos` endpoint. `APIURL` is from the `constants.js` file, which we will create later. It simply has a string with the base URL for the endpoints, to which we can make HTTP requests.

The `axios.get()` method returns a promise that resolves to an object. The object has the HTTP request. The `data` property has the response body. We assign the body data to the `this.photos` reactive property to show the photo entries in the template.

The following code is for retrieving the photo:

```
{
    ...
    methods: {
        async load() {
            const { data } = await axios.get(`${APIURL}/photos`);
            this.photos = data;
        },
        ...
    },
    beforeMount() {
        this.load();
    },
};
```

The `edit()` method calls the `this.$router.push()` method with an object that has the URL path that we want to go to. Additionally, the `path` property has the base path for the route we want to go to plus the URL parameters that we want to add to the end of the path. The `id` parameter is a URL parameter that we attach to the path. It has the ID of the photo entry:

```
{
    ...
    methods: {
        ...
        edit(id) {
```

```
            this.$router.push({ path: `/edit-photo-form/${id}` });
        },
        ...
    },
    ...
};
```

The `deletePhoto()` method also takes the `id` parameter. It is the same one as the parameter of the `edit()` method. In this method, we call the `axios.delete()` method to make a *DELETE* request to the `photos` endpoint with the `id` parameter, which was used as the URL parameter, to identify which entry to delete. Once the item has been deleted, we call the `this.load()` method to reload the latest entries from the API:

```
{
    ...
    methods: {
        ...
        async deletePhoto(id) {
            await axios.delete(`${APIURL}/photos/${id}`);
            this.photos = [];
            this.load();
        },
    },
    ...
}
```

In the `template` section, we use the `v-for` directive to render the entries of the `photos` reactive property array into a table. The `key` prop is required to identify unique items with a unique ID. The `key` prop is very important since we are going to remove items from the list when the user clicks on the **Delete** button. This means that each entry must have a unique ID so that Vue 3 can identify all of the items after we delete one item properly. This is so that the latest items can render correctly.

We render the photos using a `v-for` loop:

```
        ...
        <div v-for="p of photos" class="row" :key="p.id">
            <div>
                <img :src="p.photoFile" />
```

```
    </div>
    <div>{{p.name}}</div>
    <div>{{p.description}}</div>
    <div>
        <button @click="edit(p.id)">Edit</button>
        <button @click="deletePhoto(p.id)">Delete</button>
    </div>
  </div>
  . . .
```

To render the image, we use the `img` tag with the `src` prop. The `photoFile` property is a base64 URL that has the text form of the image. The other properties are strings that we render directly in our table. In the rightmost column, we have two buttons – **Edit** and **Delete**. The **Edit** button calls `edit()` with the `id` property of the photo entry when we click on it. This will navigate us to the photo edit form, which we will create later. The **Delete** button will call the `deletePhoto()` method with the `id` property of the photo entry to delete. The items will be reloaded once the items are deleted:

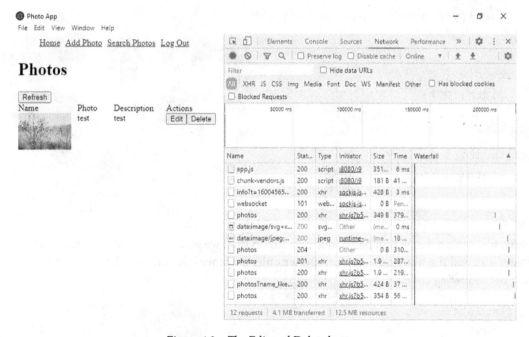

Figure 4.2 – The Edit and Delete buttons

In the `style` tag, we have some CSS to display `div` tags as a table. The `row` class has the `display` property set to `flex` so that we can use it as a `flexbox` container. The `flex-wrap` property is set to `wrap` so that we can wrap anything that overflows. Usually, we won't have anything that overflows in this template. `justify-content` is set to `space-between` in order to evenly distribute the child elements in the `flexbox` container.

The `div` tags inside the `div` tag with the `row` class have the `width` property set to `25%` so that we can evenly distribute the child elements in the row. This allows us to display the four `div` tags inside the `div` tag for the row side by side. The `img` element inside the `div` tag with the `row` class has the `width` property set to `100px` so that we can view a small thumbnail preview in the list of photos.

The `style` tag has the `scoped` attribute, which means that the styles won't affect any other components in our project:

```
<style scoped>
.row {
  display: flex;
  flex-wrap: wrap;
  justify-content: space-between;
}

.row div {
  width: 25%;
}

.row img {
  width: 100px;
}
</style>
```

Next, we create a navigation bar in our app. To do that, we go into the `src/components` folder and add the `NavBar.vue` component file. Once we have created the file, we can add the following code:

```
<template>
  <nav>
    <ul>
      <li>
        <router-link to="/">Home</router-link>
```

```
      </li>
      <li>
        <router-link to="/add-photo-form">Add Photo
          </router-link>
      </li>
      <li>
        <router-link to="/search">Search Photos
          </router-link>
      </li>
    </ul>
  </nav>
</template>
```

Here, we added a `ul` element to add an unordered list. This way, we won't see any numbers displayed on the left of each `li` element. Inside the `li` elements, we have the `router-link` component from Vue Router to display the links that allow us to navigate our app. We use `router-link` instead of a regular a tag. This is because Vue Router will resolve the `to` prop of the `router-link` component to the correct path and display the component we expect if it finds a match in the URL patterns.

Since we haven't registered the Vue Router plugin or any of the routes yet, or added `NavBar` to any component, we won't see anything in the navigation bar. The `style` tag has some styles that can make the links display horizontally instead of vertically.

Additionally, we have a **Log Out** link to log us out of the app. The `logout()` method clears the local storage with the `localStorage.clear()` method. Then, we redirect back to the login page by calling the `this.$router.push()` method with the `/login` path:

```
<script>
export default {
  methods: {
    logOut() {
      localStorage.clear();
      this.$router.push("/login");
    },
  },
};
</script>
```

The ul li selector has the list-style property set to none, so we don't see the bullet displayed to the left of the NavBar item. We display them horizontally with the display property set to inline. Then, we add the margin-right property and set it to 10px so that we have some spaces between the links.

The ul selector has the margin property set to 0 auto, so we can center the links horizontally. The width is 70vw so that they are closer to the center of the screen instead of putting the items on the left:

```
<nav>
    <ul>
      <li>
        <router-link to="/">Home</router-link>
      </li>
      <li>
        <router-link to="/add-photo-form">Add Photo
          </router-link>
      </li>
      <li>
        <router-link to="/search">Search Photos
          </router-link>
      </li>
    </ul>
  </nav>

<script>
export default {
  methods: {
    logOut() {
      localStorage.clear();
      this.$router.push("/login");
    },
  },
};
</script>

<style scoped>
ul li {
```

```
    list-style: none;
    display: inline;
    margin-right: 10px;
  }


ul {
    margin: 0 auto;
    width: 70vw;
  }
</style>...
```

Now that we have finished the form that allows us to save our photos, let's take a look at how to display the added photos on a page.

Adding a photo display

Here, we add a search page so that we can search for photo entries using their names. To do that, we add the `SearchPage.vue` component file to our project's `src/components` folder.

The `SearchPage.vue` component is simpler than the `PhotoForm` component. It has one form element with one form field. The form field is used to accept a keyword from the user to search our photo collection. The input has the `type` attribute set to `text`, so we have a regular text input in our code. As with the other inputs, we bind the input value to a reactive property with the `v-model` directive. The `id` parameter is set so that we can use the `for` attribute with the label. The form also has a **Search** button, which has an input type set to `submit`:

```
<template>
  <div>
    <h1>Search</h1>
    <form @submit.prevent="submit">
      <div>
        <label for="name">Keyword</label>
        <br />
        <input type="text" v-model="keyword" name="keyword"
          id="keyword" class="form-field" />
      </div>
      <div>
```

```
            <input type="submit" value="Search" />
        </div>
    </form>
    <div v-for="p of photos" class="row" :key="p.id">
        <div>
            <img :src="p.photoFile" />
        </div>
        <div>{{p.name}}</div>
        <div>{{p.description}}</div>
    </div>
  </div>
</template>
```

Then, the search results are displayed in a `row` class so that the items are in rows. This is similar to how we display the photos in the `HomePage` component. `img` has the base64 URL set as the value of the `src` prop. Additionally, we have the `name` and `description` properties to the right of it. The `v-for` directive loops through the `photos` reactive property array to enable us to display the data. Once again, we have the `key` prop set to a unique ID to display the items by their IDs.

In the `component options` object, we use the `data()` method to initialize our reactive properties. They include `keyword` and `photos`. The `keyword` reactive property is used for the search keyword. The `photos` reactive property is used to store the photo collection search results:

```
<script>
import axios from "axios";
import { APIURL } from "../constant";

export default {
  name: "SearchPage",
  data() {
    return {
      keyword: "",
      photos: [],
    };
  },
  ...
  watch: {
```

```
    $route: {
        immediate: true,
        handler() {
            this.keyword = this.$route.query.q;
            this.search();
        },
      },
    },
  };
</script>
```

In the `methods` property, we have a few methods that we can use. The `search()` method allows us to get the data with the `axios.get()` method. This method makes a *GET* request with a query string, so we can get the entries we are looking for. The `this.$route.query.q` property is used to get the q query parameter from the URL. This property is made available because we will register the Vue Router plugin so that we can get the `query` parameter from this property. Once we get the response data, we assign it to the `this.photos` reactive property.

The `submit()` method is run when the form is being submitted, either by clicking on the **Search** button or pressing *Enter*. Since we listen to the `submit` event in the form, this method will be run. Like with all the other forms, we add the `prevent` modifier to the `@submit` directive. This is so that we can call to the `event.preventDefault()` method to prevent data from being submitted to the server side. In this method, we call the `this.$router.push()` method to redirect the page to the `/search` path with a query string. The `/search` path will be mapped to the current component, so we just remount this component with the new query string in the URL. This way, we can set the `this.$router.query. q` property to get the query string parameter with the key to get the query string value and use it.

The `name_like` URL query parameter will be picked up by the API so that we can search for the text that we set as the value in the `name` field.

Finally, we have a watcher for the `$route` reactive property. We need to set the `immediate` property to `true` so that we get the latest value of the `query` parameter, and then run the `search()` method to get the data from the *GET* request when this component mounts. The `handler()` method has a method that runs when the `$route` object changes. The changes to the `query` property will be picked up. Therefore, inside the method, we set the `keyword` reactive property to the value of `this.$route.query.q` to display the latest value of the q query string in the input box. Additionally, we call the `this.search()` method to get the latest search results based on the query string.

The `styles` tag has some styles that we can use to style our form and rows. They are similar to the ones we had before. We make the form fields wide and display the form closer to the center. The rows are displayed with a flexbox container with all of the cells having an even width within the rows:

```
<style scoped>
.form-field {
  width: 100%;
}
.row {
  display: flex;
  flex-wrap: wrap;
  justify-content: space-between;
}

.row div {
  width: 25%;
}

.row img {
  width: 100px;
}
</style>
```

Finally, we need to create a file to export an `APIURL` variable so that the components can reference them. We have used these in most of the components we have created so far. In the `src` folder, we create the `constants.js` file and write the following code to export the `APIURL` variable:

```
export const APIURL = 'http://localhost:3000'
```

Now, we can import `SearchPage.vue` to all of our components properly and add a search page.

Adding routing to the Photo Manager app

Without the Vue Router plugin, we cannot display the page components inside our app. The links won't work and we cannot redirect anywhere. To add the Vue Router plugin, we need to register it and then add the routes. We add the following code to the `src/main.js` file:

```
import { createApp } from 'vue'
import App from './App.vue'
import { createRouter, createWebHistory } from 'vue-router'
import PhotoFormPage from './components/PhotoFormPage';
import SearchPage from './components/SearchPage';
import HomePage from './components/HomePage';

const routes = [
  { path: '/add-photo-form', component: PhotoFormPage },
  { path: '/edit-photo-form/:id', component: PhotoFormPage },
  { path: '/search', component: SearchPage },
  { path: '/', component: HomePage },
]

const router = createRouter({
  history: createWebHistory(),
  routes
})

const app = createApp(App)
app.use(router);
app.mount('#app') ...
```

In this file, we import all of the page components and then put them into the `routes` array. The `routes` array has the routes. Each object in the array has the `path` and `component` properties. The path has the URL patterns we want to match to the component and the `component` property has the `component` object that we want to load when the URL pattern matches what we have in the path. The path is a string with a URL pattern. We have one URL parameter placeholder in our string. The `:id` string has the URL placeholder for the `id` URL parameter. In our `EditPhotoFormPage` component, we retrieve the `id` URL parameter by using the `this.$route.params.id` property. It will be returned as a string.

The `createRouter()` function enables us to create a router object that we can register in our app using the `app.use()` method. This is new to Vue Router 4 and is different from Vue Router 3. The way we register the Vue Router plugin and the routes is different from Vue Router 3. Therefore, Vue Router 4 is the only version that can be used with Vue 3. The `createWebHistory()` function lets us use HTML5 mode. Using this, we can remove the hash sign between the base URL segment and the rest of the URL. This makes the URLs look better and more familiar to the user. The `routes` property has the array of routes that we created earlier. Then, to register the routes and the Vue Router plugin, we call `app.use(router)` to register both. Now the `router-link` components and the redirects should work.

The `beforeEnter()` method is a per-route navigation guard. We need this method so that we can only access the pages that are available after login. In this method, we check whether the local storage item with the key logged in is `true`. Then, if that is `false`, we redirect to the login page by calling the `next()` function with the `path` property set to `login`. The `return` keyword is required before calling `next` since we don't want to run the rest of the function's code. Otherwise, we just call `next` to continue with the navigation to the destination route, which is the value of the `path` property. We also add the `beforeEnter()` method to the route objects that we want to apply:

```
const beforeEnter = (to, from, next) => {
  const loggedIn = localStorage.getItem('logged-in') ===
    'true';
  if (!loggedIn) {
    return next({ path: 'login' });
  }
  next();
}
```

Then, in `src/App.vue`, we add the `router-view` component and the `NavBar` component by writing the following code:

```
<template>
  <div id="app">
    <nav-bar v-if="!$route.fullPath.includes
      ('login')"></nav-bar>
    <router-view></router-view>
  </div>
</template>
```

```
<script>
import NavBar from "./components/NavBar.vue";

export default {
  name: "App",
  components: {
    NavBar,
  },
};
</script>

<style scoped>
#app {
  margin: 0 auto;
  width: 70vw;
}
</style>
```

We import the `NavBar.vue` component and then register it by putting it in the `components` property. Then, we add the `nav-bar` component to display the navigation bar with the `router-link` components using router links. The `router-view` component displays the component that is matched by Vue Router by comparing the URL pattern to the patterns in the `routes` array.

Now, when we click on the links or submit the forms successfully, we will see the routes loaded.

We don't have to display the `nav-bar` component when we are on the login page. Therefore, we add a check for the `$route.fullPath.includes()` method to check whether we are on the login page. The `$route.fullPath` property has the full path without the base URL of the current page.

Using our app with photo management APIs

In the previous sections, we looked at the client-side part of the photo display. To return the photos from the API, we have to add a backend API that will enable us to store and retrieve the data for our app. Since this book is mostly focused on client-side app development with Vue 3 and not server-side app development, we will use a simple API solution to store our data with JSON so that we don't have to create our own API. All of the data is stored in a flat-file database, which is entirely JSON. To do this, we use the JSON Server package. This is a package that requires no configuration, and we can get it running in only a minute. All our fields are stored as JSON object properties, so they need to be text, including images. This package is made for frontend developers who require a backend to quickly prototype our apps.

First, we run `npm i -g json-server` to install the JSON Server package. This way, we can access the JSON Server package from any folder. Once we do that, we create a `photo-api` folder to store our photo database. Then, inside the folder, we add the `db.json` file. Next, we create the `photo-api` folder, go to the folder we just created, and run `json-server --watch db.json` to run the server. Inside the folder, we add the following code:

```
{
  "photos": []
}
```

In the `db.json` files, we will have all the endpoints that we point to in our Vue 3 app. The server should be listening to port `3000`, and so, the base URL for the API is `localhost:3000`. Now, we should have access to the following API endpoints:

- `GET /photos`
- `GET /photos/1`
- `POST /photos`
- `PUT /photos/1`
- `PATCH /photos/1`
- `DELETE /photos/1`

The `GET /photos` endpoint allows us to get all the items in the `photos` JSON array. The `GET /photos/1` endpoint returns a single photo entry with an ID of 1. We can replace it with any ID. The `POST /photos` endpoint enables us to add a new entry in the `photos` JSON array. `PUT /photos/1` and `PATCH /photos/1` allow us to update a photo entry with ID 1. The `DELETE /photos` route lets us delete a photo with an ID of 1.

The *GET* request also takes a query string. In order to search the field with a given piece of text, we can make a *GET* request to a URL such as GET /photos?tname_like=foo. This enables us to search the Name field of each entry and find the photos entries with the Name field containing the text as its value.

Now, we should be able to make the requests that we have in the client-side API to get the things we want. The whole JSON array is returned as JSON so that we can render the items easily. The JSON Server package will watch for any updates to the JSON, so we will always get the latest data. Additionally, we can change the port by using the port flag. So, we can write something such as run json-server --watch db.json –port 3005 to run the JSON server in port 3005. The APIURL variable also has to change accordingly.

When the server is running, we should see something similar to the following in Command Prompt:

```
Node.js command prompt - json-server --watch db.json

c:\photo-api>json-server --watch db.json

\{^_^}/ hi!

Loading db.json
Done

Resources
http://localhost:3000/photos

Home
http://localhost:3000

Type s + enter at any time to create a snapshot of the database
Watching...

GET /photos 200 35.556 ms - 2
GET /photos 304 28.319 ms - -
GET /photos 304 17.200 ms - -
GET /photos 304 28.183 ms - -
GET /photos 200 44.668 ms - 2
GET /photos 304 34.404 ms - -
POST /photos 201 100.199 ms - -
GET /photos 200 62.028 ms - -
GET /photos?name_like=undefined 200 32.973 ms - 2
GET /photos 304 35.267 ms - -
GET /photos?name_like=undefined 304 35.065 ms - -
GET /photos?name_like=test 200 42.607 ms - -
```

Figure 4.3 – JSON Server output

Now our Electron desktop app window has the Vue 3 photo management app displayed, and we can manipulate our photo collection as we wish. We should now be able to view our app as follows:

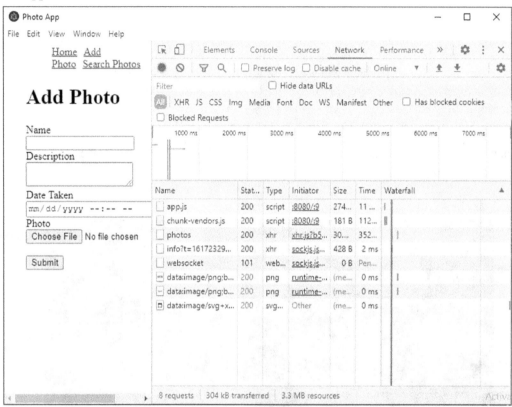

Figure 4.4 – The Photo App form

We can view the Search page in the following screenshot:

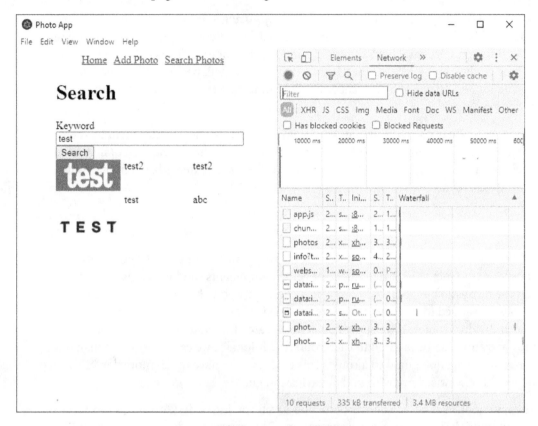

Figure 4.5 – The Search page

The photo app project is now finished. We added forms to enable users to add photo entries. We used Vue Router to allow us to add apps with multiple pages and easily navigate between them. We also added the Vue CLI Electron plugin so that we can easily build cross-platform desktop apps with Vue 3.

Summary

So, we can build apps that are beyond simple apps using a few components with Vue Router. This allows us to map different URLs to different components. This way, we can go to different URLs and have different components rendered. The URLs can have query strings, hashes, and URL parameters in them, and they will be parsed automatically by Vue Router. They will be matched with the route patterns listed in the route definition arrays.

Routes can be nested, and they can have names so that we can identify them by their name rather than their route pattern. Additionally, we can have a catchall or a 404 route to enable us to display something when none of the route patterns listed in our route definitions are matched.

Vue Router also has methods for programmatic navigation, which are named in the same way as the ones in the browser history API and can be called with the same arguments.

Additionally, we learned how to restrict some routes from being displayed unless certain conditions are met. To restrict when the route components can be seen by the user, we can use navigation guards to do our checks before we go to a route. Navigation guards can also be added to run after the route is loaded. They can also be applied globally or with individual routes. If we apply navigation guards individually, then the checks for the navigation can be tailored to each route. Additionally, we can apply more than one per-route navigation guard to a route. This enables us to have much more flexibility than we would otherwise have with global navigation guards.

Then, we looked at how to convert our Vue 3 web app into a desktop app. With Electron, we can build desktop apps from browser apps. This is very handy since we can build business apps that are web-based and convert them into desktop apps with ease. The apps are cross-platform, and we can do a lot of stuff that we can do with regular desktop apps easily. For example, we can have limited access to hardware such as cameras and microphones just as we do with browser apps. Additionally, we can show native notifications to users, as we would in a desktop app since this is supported by the Chromium browser engine. Electron just runs our app in the Chromium browser. The Vue CLI Electron Builder plugin lets us convert a Vue app into an Electron app with one command.

In the next chapter, we will build a calculator mobile app with Ionic.

5

Building a Multipurpose Calculator Mobile App with Ionic

In the first four chapters, we have built various kinds of web applications with Vue 3. We can also create mobile applications with Vue 3, but we can't create them with Vue 3 alone. We can create mobile apps with mobile app frameworks that use Vue 3 as its base framework. In *Chapter 4, Building a Photo Management Desktop App*, we built a web app with Vue Router so that we could have multiple pages in our app. Vue Router lets us create apps that are slightly complex.

In this chapter, we will move further from our knowledge of building web apps so that we can start building mobile apps. The app we will build is a calculator app that lets us convert currencies and calculate tips. It will also remember the calculations that we made so we can go back to redo them easily.

In addition, we will cover the following topics:

- Introducing Ionic Vue
- Creating our Ionic Vue mobile app project
- Installing the packages for our project
- Adding the calculators to our calculator app

Technical requirements

The source code for this chapter's project can be found at `https://github.com/PacktPublishing/-Vue.js-3-By-Example/tree/master/Chapter05`.

Introducing Ionic Vue

Ionic Vue is a mobile app framework that lets us build apps with TypeScript and Vue. It also has versions based on React and Angular. It comes with many components that we can add to our app, just like any other UI framework. They include common things such as inputs, menus, icons, lists, and much more. A compiled Ionic app runs in a web view, so we can use web technologies such as local storage, geolocation, and other browser APIs in our apps.

It also comes with built-in tools that let us build mobile apps automatically, without the need to set up everything from scratch ourselves. Ionic Vue creates components that use the Composition API by default, so we will use that to build Vue apps that are more modular and work better with TypeScript.

Understanding the Composition API

The Composition API works better with TypeScript because it does not reference the `this` keyword, which has a dynamic structure. Instead, everything that is part of the Composition API, including its libraries, are compatible with it functions, which have clear parameters and return types. This lets us define TypeScript type definitions for them easily. Vue Router 4 and Vuex 4 are compatible with Vue 3's Composition API, so we can use them together in our Ionic Vue app.

With the Composition API, we still have the component object, but its structure is completely different from what we have in the Options API. The only property that is the same between the options and the Composition API is the `components` property. They both let us register components in both APIs. Our component logic is mostly in the `setup()` method. This is where we define our reactive properties, computed properties, watchers, and methods. Third-party libraries may also provide us with hooks, which are functions that we can call in our setup function to give us functionality that we want from the library. For example, Vuex 4 gives us the `useStore` hook so that we can gain access to the store. Vue Router 4 comes with the `useRouter` hook to let us navigate in our app.

The way we define reactive and computed properties is different from the options API. In the options API, which we used in the previous chapters, we define and initialize our reactive properties in the `data()` method. In the Composition API, we call the `ref` function to define reactive properties that hold primitive values; we then call `reactive` to define reactive properties that have object values. To define computed properties in the Composition API, we call the `computed` function with a callback that references other reactive properties to create the computed property.

Watchers are created with the `watch` function. It takes a callback to return the reactive property we want to watch the value for. The second argument we pass into the `watch` function is a callback that lets us do something when the watched value changes. We can get the latest value of the reactive property being watched with the first parameter of the callback. The third argument contains the options for the watchers. We can set the deep and immediate properties in there as we do with watchers in the Options API.

Methods are also added to the `setup` function. We can use arrow functions or regular functions to define them since the value of this doesn't matter. Reactive properties and methods must be returned in the object we return in the `setup()` method to expose them to the template. This way, they can be used in our code.

Understanding TypeScript

TypeScript is a language made by Microsoft that is an extension of JavaScript. It provides us with compile-time checking for data types in our code. However, it doesn't provide us with extra runtime data type checks since TypeScript compiles to JavaScript before they are run. With the Composition API, our components do not reference the `this` keyword, so we don't have to worry about it having the wrong value.

The benefit of using TypeScript is to ensure the type safety of primitive values, objects, and variables, which do not exist within JavaScript. In JavaScript, we can assign anything to any variable. This is, of course, going to be a problem since we may assign things to data types that we wouldn't usually. Also, functions can take anything as parameters and we can pass in any argument in any order into functions, so we may run into problems there if we pass in arguments that a function doesn't expect. Also, things may become `null` or `undefined` anywhere, so we must make sure that only places where we expect things to be null or undefined have those values. JavaScript functions can also return anything, so TypeScript can also restrict that.

Another big feature of TypeScript is that we can create interfaces to restrict the structure of objects. We can specify object properties and their types so that we can restrict objects to have the given properties, and so that the properties have the data types that we specify. This prevents us from assigning objects to variables and parameters that we don't expect, and it also provides us with autocomplete functionality in text editors that support TypeScript that we can't get with JavaScript objects. This is because the structures of objects are set. Interfaces can have optional and dynamic properties to let us maintain the flexibility of JavaScript objects.

To retain the flexibility of JavaScript, TypeScript comes with union and intersection types. **Union** types are where we have more than one type joined together with a logical OR operator. A variable with a union type means that a variable can be one of the list of types that is of the union type. **Intersection** types are multiple types joined together with a logical AND operator. A variable with a type set to an intersection type must have all the members of all the types in the intersection.

To keep type specifications short, TypeScript comes with the `type` keyword, which lets us create a type alias. Type aliases can be used like regular types, so we can assign type aliases to variables, properties, parameters, return types, and so on.

In our mobile app, we will add pages for a tip calculator, currency converter, and a home page with a past list of calculations. We have any calculations that we have made in local storage so that we can go back to them later. The history is saved to local storage via the `vuex-persistedstate` plugin. This plugin is compatible with Vuex 4 and it lets us save the Vuex state to local storage directly, without us writing any extra code to do that ourselves.

Now that we've looked at the basics of Vue's Composition API, TypeScript, and Ionic, we can start building our app with it.

Creating our Ionic Vue mobile app project

We can create our Ionic Vue project by installing the Ionic CLI. First, we must install the Ionic CLI by running the following command:

```
npm install -g @ionic/cli
```

Then, we must create our Ionic Vue project by going to the folder where we want our project folder to be running. We can do this with the following command:

```
ionic start calculator-app sidemenu --type vue
```

The `sidemenu` option lets us create an Ionic project with a side menu added to its pages. This will save us time with we're creating the menus and pages. The `--type vue` option lets us create an Ionic Vue project.

We can get help with all the options and look at an explanation of each by using the following commands:

- `ionic -help`
- `ionic <command> --help`
- `ionic <command><subcommand> --help`

We should run `ionic <command> --help` in our project directory.

Using Capacitor and Genymotion

Ionic Vue projects are served and built with Capacitor. Capacitor will open the project in Android Studio; then, we can launch it from there and preview our app in an emulator or device. For this project, we will preview our app with the Genymotion emulator. It is fast and has a plugin that lets us launch from Android Studio. We can download Genymotion emulator from `https://www.genymotion.com/download/` and Android Studio can be downloaded from `https://developer.android.com/studio`.

Once we've installed Genymotion, we must create a virtual machine from the Genymotion UI. To do this, we can click the *plus* button, and then add the device that we want. We should add a device that has a recent version of Android, such as Android 7 or later. The other options can be chosen according to our preference. To install the Genymotion plugin for Android Studio, follow the instructions at `https://www.genymotion.com/plugins/`. This will let us run our Android Studio project in Genymotion.

Next, in the `package.json` file in our project, if we don't see the `ionic:serve` and `ionic:build` scripts in the scripts section, we can add them by writing the following code inside the scripts section of our `package.json` file:

```
"ionic:serve": "vue-cli-service serve",
"ionic:build": "vue-cli-service build"
```

Then, we can run `ionic build` to build our code so that it can be served with Capacitor later.

Once we've done that, we can run the following command to add the dependencies for an Android project:

```
npx cap add android
npx cap sync
```

This is also required so that we can run our project as an Android app.

Once we've run those commands, we can run the following command so that we can run our app with live reload and make network access available from Genymotion:

```
ionic capacitor run android --livereload --external
--address=0.0.0.0
```

This way, we can access the internet just like any other app can. It also runs the `ionic:serve` script so that we can preview our app in the browser. Previewing our app in the browser is faster than in the emulator, so we may want to do that:

Figure 5.1 – Genymotion emulator

If we want to preview in Genymotion, we can go to Android Studio, which should open automatically once we run the `ionic capacitor run` command. Then, we can press *Alt+Shift+F10* to open the run app dialog, and then choose the app to run it.

Now that we have set up our Vue Ionic project, we must install a few more packages so that we can create our mobile app.

Installing the packages for our project

We must install some dependencies that are needed in the project, but they aren't installed yet. We can use Axios to make HTTP requests to get the exchange rate. The `uuid` module lets us generate unique IDs for our history entries. Vuex doesn't come with the Ionic project, so we have to install that. We must also install the `vuex-persistedstate` module so that we can save Vuex state data to local storage.

To install these packages, run the following command:

```
npm install axios uuid vuex@next vuex-persistedstate
```

The next version of Vuex is the 4.x version, which is compatible with Vue 3.

Adding the calculators to our calculator app

Now that we have our project ready, we can start working on our app. We start with adding the route definition to map URL paths to the page components that we will create. Then we will work on the components for each feature. And then we will add the Vuex store with code to persist the store data to local storage so we can use the data whenever we want.

Adding routes

First, we will work on adding routing to our calculator app. In the `src/router/index.ts` file, write the following code:

```
import { createRouter, createWebHistory } from '@ionic/vue-
  router';
import { RouteRecordRaw } from 'vue-router';

const routes: Array<RouteRecordRaw> = [
  {
    path: '/',
    component: () => import('../views/Home.vue')
  },
  {
    path: '/currency-converter',
    component: () =>
      import('../views/CurrencyConverter.vue')
  },
  {
    path: '/tips-calculator',
    component: () => import('../views/TipsCalculator.vue')
  }
]
```

```
const router = createRouter({
  history: createWebHistory(process.env.BASE_URL),
  routes
})


export default router
```

In this file, we have the `routes` array, which we can use to add the routes for the pages that we are going to add to our calculator app. The `routes` array is of the `Array<RouteRecordRaw>` type. This means that the objects in the `routes` array must have the path and component properties. The `path` property must be a string, while the component can be a component or function that returns a promise that resolves to a component. If the objects don't match the structure specified by `Array<RouteRecordRaw>`, the TypeScript compiler will give us an error when we build the code.

The code is built whenever we change any code file since we have the `livereload` option set, so we will get compiler errors almost immediately. This prevents most data type-related errors from occurring during runtime. The type definitions are built into the `vue-router` module, so we don't have to worry about missing data types.

Adding the currency converter page

Next, we will add the currency converter page. To add it, first, create the `src/views/CurrencyConverter.vue` file. Then, we must add the header to the template by writing the following code:

```
<template>
  <ion-page>
    <ion-header translucent>
      <ion-toolbar>
        <ion-buttons slot="start">
          <ion-menu-button></ion-menu-button>
        </ion-buttons>
        <ion-title>Currency Converter</ion-title>
      </ion-toolbar>
    </ion-header>

    ...

  </ion-page>
</template>
```

The `ion-page` component is the page container that lets us add content inside it. The `ion-toolbar` component adds a toolbar to the top of the page. The `ion-buttons` component is a container for buttons and inside it, we must add the `ion-menu-button` to the start slot so that we can add a menu button to the top-left corner of the screen. The `ion-menu- button` component will open the left-hand side menu when we click it. The `ion-title` component contains the page title. It is located at the top-left corner.

Next, we must add the `ion-content` component to add the content to the currency converter page. For instance, we can write the following code:

```
<template>
  <ion-page>
    . . .
    <ion-content fullscreen>
      <div id="container">
        <ion-list>
          <ion-item>
            <ion-label :color="!amountValid ? 'danger' :
              undefined">
              Amount to Convert
            </ion-label>
            <ion-input
              class="ion-text-right"
              type="number"
              v-model="amount"
              required
              placeholder="Amount"
            ></ion-input>
          </ion-item>
          . . .
        </ion-list>
        . . .
      </div>
    </ion-content>
  </ion-page>
</template>
```

Here, we added the `ion-list` component so that we can add a list to our page. It lets us add a list of items to our app. In `ion-list`, we add `ion-item` to add a list item component. `ion-label` lets us add the label into the list item. The `color` property of the label text is set by the `color` prop. The `amountValid` prop is a computed property that checks whether the `amount` reactive property is valid. The `ion-input` component renders an input into our app. We set `type` to `number` to make the input a numeric input.

The `placeholder` prop lets us add a placeholder to our app. The `ion-text-right` class lets us put the input on the right-hand side. This is a class that comes with the Ionic framework. The `v-model` directive lets us bind the `amount` reactive property to the inputted value so that we can use the inputted value in the component code.

The `fullscreen` prop of `ion-content` makes the page full screen.

Next, we will add more items to the `ion-list` component:

```
<template>
  <ion-page>
. . .

    . . .
    <ion-content fullscreen>
      <div id="container">
        <ion-list>


          . . .
          <ion-item>
            <ion-label> Currency to Convert From
            </ion-label>
            <ion-select
              v-model="fromCurrency"
              ok-text="Okay"
              cancel-text="Dismiss"
            >
              <ion-select-option
                :value="c.abbreviation"
                v-for="c of fromCurrencies"
                :key="c.name"
              >
```

```
                {{ c.name }}
            </ion-select-option>
          </ion-select>
        </ion-item>
        ...
      </ion-list>
      ...
    </div>
  </ion-content>
  </ion-page>
</template>
```

Here, we have added more `ion-items` to our `ion-list`. The `ion-select` component lets us add the currency to convert from dropdown, which lets us choose the currency that the amount is in. We bind `fromCurrency` to the value we selected in the dropdown to get the selected item inside our component code. The `ok-text` prop sets the OK text in the dropdown, while the `cancel-text` prop contains the text for the cancel button. The `ion-select` component lets us show a dialog with radio buttons that lets us display the items for us to choose from. Then, when we click or tap on the **OK** button, we can select the item.

The `ion-select-option` component lets us add options to the select dialog box. We use the `v-for` directive to loop through the `fromCurrencies` reactive property, which is a computed property that we create from filtering out the `selected` option from the **Currency to Convert To** dialog, which we will add later. This way, we can't select the same currency in both dropdowns, so currency conversion makes sense.

Next, we will add another select dialog to let us select the currency that we want to convert the amount into. To do this, we can write the following code:

```
<template>
  <ion-page>
    ...
    <ion-content fullscreen>
      <div id="container">
        <ion-list>
          ...
  ...
                {{ c.name }}
```

```
            </ion-select-option>
          </ion-select>
        </ion-item>

        <ion-item>
          <ion-button size="default"
            @click.stop="calculate">
            Calculate
          </ion-button>
        </ion-item>
      </ion-list>

      ...
    </div>
  </ion-content>
  </ion-page>
</template>
```

The toCurrencies reactive property is a computed property that contains an entry that has the value of fromCurrency filtered out. This means we can't select the same currency in both dropdowns.

We also added the **Calculate** button, which lets us calculate the converted amount. We will add the calculate() method shortly.

Next, we will add another ion-list. This will add a list that adds the labels to display the converted amount. To do this, we can write the following code:

```
<template>
  <ion-page>
    <ion-header translucent>
      <ion-toolbar>
        <ion-buttons slot="start">
          <ion-menu-button></ion-menu-button>
        </ion-buttons>
        <ion-title>Currency Converter</ion-title>
      </ion-toolbar>
    </ion-header>

    <ion-content fullscreen>
```

```
        <div id="container">
          . . .
          <ion-list v-if="result">
            <ion-item>
              <ion-label>Result</ion-label>
              <ion-label class="ion-text-right">
                {{ amount }} {{ fromCurrency }} is {{
                  result.toFixed(2) }}
                {{ toCurrency }}
              </ion-label>
            </ion-item>
          </ion-list>
        </div>
      </ion-content>
    </ion-page>
</template>
```

Here, we have displayed the amount and `fromCurrency` that we entered. We also
displayed the result and the `toCurrency` option that we selected. We called `toFixed`
with argument 2 to round the result to two decimal places.

Adding the script tag

Next, we will add a `script` tag with the `lang` attribute set to `ts` so that we can add
the TypeScript code. First, we will add the `import` statements in order to add the
components and other items that we will use in our code:

```
<script lang="ts">
import {
  IonButtons,
  IonContent,
  IonHeader,
  . . .
} from "@ionic/vue";
import { computed, reactive, ref, watch } from "vue";
import { currencies as currenciesArray } from "../constants";
import axios from "axios";
import { useStore } from "vuex";
```

```
import { CurrencyConversion } from "@/interfaces";
import { v4 as uuidv4 } from "uuid";
import { useRoute } from "vue-router";

...
</script>
```

See this book's GitHub repository for the full list of components that can be registered.

The computed function lets us create the computed properties that we can use with the Composition API. The reactive function lets us create reactive properties that have objects as values. The ref property lets us create computed properties that have primitive values. The watch function lets us create watchers that can be used with the Composition API.

The currenciesArray variable is an array of currencies that we will use to create the fromCurrencies and toCurrencies computed properties. The axios object lets us use the Axios HTTP client to make HTTP requests. The useStore variable is a function that lets us get access to our Vuex store. The CurrencyConversion interface provides the interface that we use to restrict the structure for the object that we add to the history list. The uuidv4 variable is a function that lets us create UUIDs, which are unique IDs that we assign to the history entries to identify them. The useRoute function lets us access the route object to get the current URL path and other parts of the URL.

Next, we will register the components by adding the components property and the component we imported into it. To do this, we can write the following code:

```
<script lang="ts">
...
export default {
  name: "CurrencyConverter",
  components: {
    IonButtons,
    ...
  },
  ...
};
</script>
```

See this book's GitHub repository for the full list of components that can be registered. We just put all the `import` components into the `component` property to register them.

Working on the setup method

Next, we will start working on the `setup()` method and add the reactive and computed properties to it. We will also add the watchers, which let us watch for route changes. First, we will write the following code:

```ts
<script lang="ts">
...
export default {
  ...
  setup() {
    const amount = ref(0);
    const fromCurrency = ref("USD");
    const toCurrency = ref("CAD");
    const result = ref(0);
    const currencies = reactive(currenciesArray);
    const store = useStore();
    const route = useRoute();

    ...
    return {
      amount,
      fromCurrency,
      toCurrency,
      currencies,
      fromCurrencies,
      toCurrencies,
      amountValid,
      calculate,
      result,
      addToHistory,
    };
  },
};
</script>
```

We call the `useStore()` method to return the store object, which contains the Vuex store. We need the Vuex store to commit mutations, which lets us add entries to our history. Because we will add the `vuex-persistsedstate` plugin to our Vuex store, the history entries will be added to local storage automatically. Similarly, we call the `useRoute` function to return the route object, which lets us get access to the route object. We need the route object to let us watch the query string for the `id query` parameter. If we find an item by their ID, then we can set the `fromCurrency`, `toCurrency`, and `amount` values by using their values from the Vuex store, which we get from local storage.

Also, we call the `ref` function to create the `amount` reactive properties, which are number values. The `fromCurrency` and `toCurrency` reactive properties are string values and they contain the currency code of the currency that we choose. The `currencies` reactive property is a reactive array that is set to `currenciesArray` as its initial value. The arguments that we pass into `ref` and `reactive` are the initial values for each reactive property.

Next, we can add the computed properties by writing the following code:

```ts
<script lang="ts">
...
export default {
  ...
  setup() {
    ...
    const fromCurrencies = computed(() => {
      return currencies.filter(
        ({ abbreviation }) => abbreviation !==
          toCurrency.value
      );
    });
...
    return {
      amount,
      fromCurrency,
      toCurrency,
      currencies,
      fromCurrencies,
      toCurrencies,
      amountValid,
```

```
        calculate,
        result,
        addToHistory,
    };  },
};
</script>
```

We call the `computed` function with a callback to create the computed property. Like with the options API, we return the value that we want for the computed property. The only thing that's different is that we get the value of a primitive value reactive property with the `value` property. The `fromCurrencies` reactive property is created by filtering the currencies entry with the abbreviation that has the same value as `toCurrency`. `toCurrencies` is created by filtering the currencies entry with the abbreviation value, which is the same as the value of `fromCurrency`.

The `amountValid` computed property lets us determine whether the amount that's entered inside `ion-input` is valid. We want it to be a number that's at least 0, so we return that condition so that we can check for this.

Next, we will add these methods to our `CurrencyConverter` component by adding more items to the `setup()` method:

```
<script lang="ts">
...
export default {
  ...
  setup() {
    ...
    const addToHistory = (entry: CurrencyConversion) =>
      store.commit("addToHistory", entry);

    const calculate = async () => {
      result.value = 0;
      if (!amountValid.value) {
        return;
      ...
    });
    result.value = amount.value *
```

```
        rates [toCurrency.value] ;
    };
    ...
    return {
        amount,
        fromCurrency,
        toCurrency,
        currencies,
        fromCurrencies,
        toCurrencies,
        amountValid,
        calculate,
        result,
        addToHistory,
    };  },
};
</script>
```

The addToHistory() method lets us add a history entry to our Vuex store and local storage so that we can show the activities on the **Home** page. This way, we can choose them later and do the same calculation. In the signature, we annotate the type of the entry parameter with the CurrencyConversion interface so that we know we are adding the right thing to the Vuex store and local storage. We commit addToHistory to the store with the history entry as the payload.

Working on the calculate method

In the calculate() method, we reset the value of result to 0. Then, we call addToHistory to add the entry to the history. The id property is generated from the uuidv4 function to generate a unique ID for the entry. We set the other properties from the reactive property values. The value property is required to access primitive value reactive properties.

Then, we use Axios to get the exchange rate from the free to use Exchange Rate API. We just have to set the base query parameter to the code of the currency that we are converting from. Finally, we compute the result of the converted value by multiplying the amount by the exchange rate retrieved from the API.

Then, to finish off the `CurrencyConverter` component, we add the watcher for the query string. We watch the `queryID` parameter, which will change if we open the history entry from the home page. To add the watcher, we can write the following code:

```ts
<script lang="ts">
...
export default {
  ...
  setup() {
    ...
    watch(
      () => route.query,
      (query) => {
        const { id: queryId } = query;
        const { history } = store.state;
        const entry = history.find(
          ({ id }: CurrencyConversion) => id === queryId
        );
        if (!entry) {
          return;
        }
      ...
      fromCurrency,
      toCurrency,
      currencies,
      fromCurrencies,
      toCurrencies,
      amountValid,
      calculate,
      result,
      addToHistory,
    };
  },
};
</script>
```

To create the watcher, we pass in a function that returns `route.query` to return the query object. The `route` variable is assigned to the returned value of the `useRoute` function that we called earlier. Then, we get the query object's value from the first parameter of the function in the second argument. We get the `id` property from the query object. Then, we get the history state from the store, which contains all the entries we stored in local storage. Local storage is automatically synchronized by `vuex-persistedstate` to the Vuex store.

We call the `history.find()` method to find the entry by its `id`. Then, an entry is returned, and we set the `retrieved` property values to the reactive properties values. This automatically populates them when we choose an entry from the history.

In the third argument, we have an object that has the immediate property set to `true` so that the watcher runs immediately when the component is being mounted.

We return everything that we want to expose to the templates with the `return` statement at the end. We include all the reactive properties, computed properties, and methods so that they can be used in templates.

Once we're finished with the project, the currency converter should look as follows:

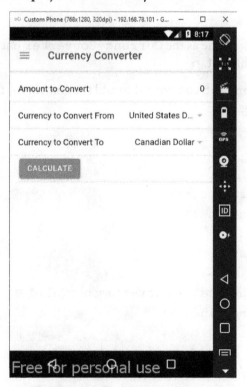

Figure 5.2 – Currency Converter

Adding the Tips calculator

Next, we will add the `TipsCalculator` page component. To add it, we must add the `src/views/TipCalculator.vue` file. Inside it, we will start by adding the template and a header:

```
<template>
  <ion-page>
    <ion-header translucent>
      <ion-toolbar>
        <ion-buttons slot="start">
          <ion-menu-button></ion-menu-button>
        </ion-buttons>
        <ion-title>Tips Calculator</ion-title>
      </ion-toolbar>
    </ion-header>
    ...
  </ion-page>
</template>
```

`ion-header` is almost the same as the `CurrencyConverter` one, except the `ion-title` content is different.

Next, we add the `ion-content` component to add the content for the page. To do this, we can write the following code:

```
<template>
  <ion-page>
    ...
    <ion-content fullscreen>
      <div id="container">
        <form>
          <ion-list>
            <ion-item>
              <ion-label :color="!amountValid ? 'danger' :
                undefined">
                ...
              {{ c.name }}
              </ion-select-option>
            </ion-select>
```

```
            </ion-item>
      ...
        </ion-list>
      ...
      </form>
    </div>
   </ion-content>
  </ion-page>
</template>
```

In the preceding code, we added `ion-list` and the ion items for the form controls. We have one that lets us enter the subtotal on the page. This is the amount before the tip. The second `ion-item` component lets us add the `country` `ion-select` control. It lets us choose a country so that we can get the tip rate for that country. The tipping rate is computed from the `tippingRate` computed property. `ion-select-option` is created from the `countries` reactive array property, which provides the list of countries that we can choose from to get their tipping rates.

Next, we will add the display for the tipping rate and the **Calculate Tip** button. To do this, we will write the following code:

```
<template>
  <ion-page>
    ...
    <ion-content fullscreen>
      <div id="container">
        <form>
          <ion-list>
            ...
            <ion-item>
              <ion-label>Tipping Rate</ion-label>
              <ion-label class="ion-text-right">{{"> {{
                tippingRate }}% </ion-label>
            </ion-item>

            <ion-item>
              <ion-button size="default"
                @click="calculateTip">
```

```
                      Calculate Tip
                  </ion-button>
              </ion-item>
          </ion-list>
...
          ...
        </form>
      </div>
    </ion-content>
  </ion-page>
</template>
```

We just display the `tippingRate` computed property's value and the **Calculate Tip** button. We add a click handler by adding the `@click` directive and setting that to the `calculateTip()` method.

The final part of the template is the calculated results. We add `ion-list` to component to add the results. We display the tip and the subtotal added together. To add it, we can write the following code:

```
<template>
  <ion-page>
    ...
    <ion-content fullscreen>
      <div id="container">
        <form>
          ...
          <ion-list>
            <ion-item>
              <ion-label>Tip (Local Currency)</ion-label>
              <ion-label class="ion-text-right">{{"> {{
                result.tip }} </ion-label>
            </ion-item>

            <ion-item>
              <ion-label>Total (Local Currency)</ion-label>
              <ion-label class="ion-text-right">{{"> {{
                result.total }} </ion-label>
```

```
          </ion-item>
        </ion-list>
      </form>
    </div>
  </ion-content>
  </ion-page>
</template>
```

Next, we will add the TypeScript code for the `TipsCalculator` component. Its structure is similar to the `CurrencyConverter` component. First, we add the imports by writing the following code:

```
<script lang="ts">
import {
  IonButtons,
  IonContent,
  IonHeader,
  IonMenuButton,
  IonPage,
  IonTitle,
  IonToolbar,
  IonSelect,
  IonSelectOption,
  IonInput,
  IonLabel,
  IonButton,
  IonList,
  IonItem,
} from "@ionic/vue";
import { computed, reactive, ref, watch } from "vue";
import { countries as countriesArray } from "../constants";
import { useStore } from "vuex";
import { TipCalculation } from "@/interfaces";
import { v4 as uuidv4 } from "uuid";
import { useRoute } from "vue-router";

...
</script>
```

We import all the components and libraries as we did with `CurrencyConverter.vue`.

Then, we register the components as we did with `CurrencyConverter`:

```ts
<script lang="ts">
...
export default {
  name: "TipsCalculator",
  components: {
    IonButtons,
    IonContent,
    IonHeader,
    IonMenuButton,
    IonPage,
    IonTitle,
    IonToolbar,
    IonSelect,
    IonSelectOption,
    IonInput,
    IonLabel,
    IonButton,
    IonList,
    IonItem,
  },
  ...
};
</script>
```

Then, we define the reactive properties and get the store and route within the `setup()` method:

```ts
<script lang="ts">
...
export default {
...
  ...
  setup() {
    const store = useStore();
```

```
    const route = useRoute();
    const subtotal = ref(0);
    const countries = reactive(countriesArray);
    const country = ref("Afghanistan");
    ...
    return {
      subtotal,
      country,
      countries,
      tippingRate,
      amountValid,
      result,
      calculateTip,
    };
  },
};
</script>
```

We call useStore and useRoute as we did within CurrencyConverter. Then, we create the subtotal reactive property with the ref function. Since its value is a number, we use the ref function to create it. The country array's reactive property is created with the reactive function.

Next, we must add some computed properties by writing the following code:

```
<script lang="ts">
...
export default {
  ...
  setup() {
    ...
    const tippingRate = computed(() => {
      if (["United States"].includes(country.value)) {
        return 20;
      } else if (
        ["Canada", "Jordan", "Morocco", "South
          Africa"].includes(country.value)
      ) {
```

```
      return 15;
    } else if (["Germany", "Ireland",
      "Portugal"].includes(country.value)) {
      return 5;
    }
    return 0;
  });
  const amountValid = computed(() => {
    return +subtotal.value >= 0;
  });
  ...
  return {
    subtotal,
    country,
    countries,
    tippingRate,
    amountValid,
    result,
    calculateTip,
  };
  },
};
</script>
```

Here, we computed the tipping rate according to the country we selected.

The `amountValid` computed property lets us check whether the `subtotal` value is valid. We want to it to be 0 or bigger.

Next, we will add the rest of the items to the component:

```
<script lang="ts">
...
export default {
  ...
  setup() {
    ...
    const result = reactive({
```

```
        tip: 0,
        total: 0,
    });

    const addToHistory = (entry: TipCalculation) =>
        store.commit("addToHistory", entry);

        ...

        tippingRate,
        amountValid,
        result,
        calculateTip,
    };
  },
};
</script>
```

The `result` reactive property contains the result of the tip calculation. The `tip` property contains the tip amount. Finally, the `total` property contains the total of both `subtotal` and `tip`.

The `calculateTip()` method lets us calculate the tip. The value of the `result` property is initiated to 0 first. Then, we check if `amountValid` is true. If it's not, we stop running the function. Otherwise, we add the history entry to the store and local storage with the `addToHistory` function. Next, we do the tip calculation with the last two lines of the `calculateTip()` method.

Finally, we add the watcher to the `setup()` method by writing the following code:

```
<script lang="ts">
...
export default {
...
  setup() {
    ...
    watch(
        () => route.query,
        (query) => {
            const { id: queryId } = query;
            const { history } = store.state;
```

```
        const entry = history.find(({ id }: TipCalculation)
          => id === queryId);
        if (!entry) {
          return;
        }
        const {
          subtotal: querySubtotal,
          country: queryCountry,
        }: TipCalculation = entry;
        subtotal.value = querySubtotal;
        country.value = queryCountry;
      },
      { immediate: true }
    );
    return {
      subtotal,
      country,
      countries,
      tippingRate,
      amountValid,
      result,
      calculateTip,
    };
  },
};
</script>
```

Just like in `CurrencyConverter.vue`, we watch the parsed query string object and populate the value from the history entry if it is found.

And finally, we return all the items we want to expose to the template, including any reactive and computed properties and the method with the `return` statement. Once we've finished the project, we should see the following screen:

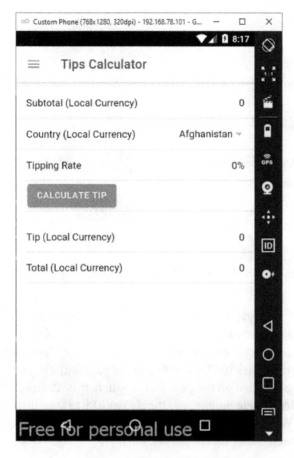

Figure 5.3 – Tips Calculator

Adding the Home page

Next, we will add the Home . vue page component, which will let us view the calculations that we have done so far. We can redo the calculation by opening the page for the calculation with the values populated from our history. To add the calculation history list, we will start with its template:

```
<template>
  <ion-page>
    <ion-header translucent>
      <ion-toolbar>
        <ion-buttons slot="start">
          <ion-menu-button></ion-menu-button>
```

```
      </ion-buttons>
        <ion-title>Home</ion-title>
    </ion-toolbar>
  </ion-header>

  <ion-content fullscreen>
    ...
        </ion-list>
      </div>
    </ion-content>
  </ion-page>
</template>
```

We have used the same header that we've used on the other pages, but this one has a different title.

Then, we render the `historyWithTypes` computed property to render the items from the history. If the `type` property is set to `tip`, we render the tip calculation data. Otherwise, we show the currency conversion data. In each row, we have the **Open** button, which calls the `go()` method when we click it. This takes us to the page with the given values from the history that have been populated on the page by the watchers of `CurrencyCoverter` or `TipsCalculator`. The **Delete** button calls the `deleteEntry()` method, which deletes the entry by its index. We must remember to set the `key` prop to the unique ID for each entry so that Vue can keep track of them properly.

Next, we will add the imports by writing the following code:

```
<script lang="ts">
import {
  IonButtons,
  IonContent,
  IonHeader,
  IonMenuButton,
  IonPage,
  IonTitle,
  IonToolbar,
  IonLabel,
  IonButton,
```

```
    IonList,
    IonItem,
} from "@ionic/vue";
import { useStore } from "vuex";
import { computed } from "vue";
import { CurrencyConversion, TipCalculation } from
  "@/interfaces";
import { useRouter } from "vue-router";
...
</script>
```

Then, we will add the `type` alias for our history entries and register the component code by writing the following code:

```
<script lang="ts">
...
type HistoryEntry = CurrencyConversion | TipCalculation;

export default {
  name: "Home",
  components: {
    IonButtons,
    IonContent,
    IonHeader,
    IonMenuButton,
    IonPage,
    IonTitle,
    IonToolbar,
    IonLabel,
    IonButton,
    IonList,
    IonItem,
  },
  ...
};
</script>
```

We create the `HistoryEntry` TypeScript type alias, which is the union of the `CurrencyConversion` and `TipCalculation` interfaces. Objects of the `HistoryEntry` type must have the structure of either the `CurrencyConversion` or `TipCalculation` interface. Then, we register the components just like we registered the other components.

Next, we will add the `setup()` method to add the component's logic:

```ts
<script lang="ts">
...
export default {
  ...
  setup() {
    const store = useStore();
    const router = useRouter();
    const history = computed(() => store.state.history);
    const historyWithTypes = computed(() => {
      return history.value.map((history: HistoryEntry):
        HistoryEntry & {
        type: string;
      } => {
        if ("subtotal" in history) {
          return {
            ...history,
            type: "tip",
          };
        }
        return {
          ...history,
          type: "conversion",
        };
      });
    });

    const go = (history: HistoryEntry & { type: string })
      => {
      const { type, id } = history;
      if (type === "tip") {
```

```
        router.push({ path: "/tips-calculator", query: { id
          } });
      } else {
        router.push({ path: "/currency-converter", query: {
          id } });
      }
    };


    const deleteEntry = (index: number) => {
      store.commit("removeHistoryEntry", index);
    };


    return {
      history,
      historyWithTypes,
      go,
      deleteEntry,
    };
  },
};
</script>
```

We get the store and the router as usual with `useStore` and `useRouter`, respectively. Then, we get the history state from the Vuex store with the `history` computed property. Then, we use the `history` computed property to create the `historyWithTypes` computed property. This lets us add the `type` property to the object so that we can distinguish the types of items in the template easily. In the `map` callback, we set the return type to `HistoryEntry & { type: string }` to create an intersection type with the interfaces that `HistoryEntry` is composed of and the `{ type: string }` type. `HistoryEntry & { type: string }` is the same as `CurrencyConversion & { type: string } | TipCalculation & { type: string }` since the `&` operator distributes when it's used with the union (`|`) operator.

The `go()` method lets us navigate to the right page with the `id` property as the value of the `id` query parameter when we call `router.push`. The `path` property contains the URL path we specified in the route definitions, while the `query` property contains the object that is used to form the query string after the path.

The `deleteEntry()` method lets us delete an entry by committing the `removeHistoryEntry` mutation to do so.

We return all the methods and computed properties so that they can be used in the template. The **Home** page should look as shown in the following screenshot:

Figure 5.4 – Home screen

Creating the Vuex store

Now, we need to create the Vuex store. To do this, we will create the `src/vue/index.ts` file and write the following code:

```
import { createStore } from 'vuex'
import createPersistedState from "vuex-persistedstate";
```

```
import {
  CurrencyConversion,
  TipCalculation
} from '../interfaces'

type HistoryEntry = CurrencyConversion | TipCalculation

const store = createStore({
  plugins: [createPersistedState()],
  state() {
    return {
      history: []
    }
  },
  mutations: {
    addToHistory(state: { history: HistoryEntry[] }, entry:
      HistoryEntry) {
      state.history.push(entry)
    },
    removeHistoryEntry(state: { history: HistoryEntry[] },
      index: number) {
      state.history.splice(index, 1)
    },
  }
})

export default store
```

Here, we have the interfaces and the same type alias as Home.vue. We created the Vuex store with the createStore function. The plugins property is set to the array that the createPersistedState function returns to let us save the store state to local storage. We have the history state in the state() method. The mutations() method has the addToHistory mutation, which lets us add entries to the history array state. We also have removeHistoryEntry, which lets us remove history items from the history state by its index. We must remember to export the store at the end so that we can import it later.

Then, we need to add the list of countries and currencies. To add them, we will create the `src/constants.ts` file and add the following code:

```
import { Choice } from "./interfaces";

export const countries: Choice [] = [
  {
    "name": "Afghanistan",
    "abbreviation": "AF"
  },
  {
    "name ": "Åland Islands",
    "abbreviation": "AX"
  },
  ...
]


export const currencies: Choice[] = [
  {
    "name": "United States Dollar",
    "abbreviation": "USD"
  },
  {
    "name": "Canadian Dollar",
    "abbreviation": "CAD"
  },
  {
    "name": "Euro",
    "abbreviation": "EUR"
  },
]
```

The full file's content can be found at `https://github.com/PacktPublishing/-Vue.js-3-By-Example/blob/master/Chapter05/src/constants.ts`.

Now, we will add the interfaces we imported by adding the `src/interfaces.ts` file and add the following code:

```
export interface Choice {
  name: string,
  abbreviation: string
}

export interface CurrencyConversion {
  id: string,
  amount: number,
  fromCurrency: string,
  toCurrency: string,
}

export interface TipCalculation {
  id: string,
  subtotal: number,
  country: string,
}
```

In `main.ts`, we must add the store to our app by writing the following code:

```
...

const app = createApp(App)
  .use(IonicVue)
  .use(router)
  .use(store);

...
```

We added `.use(store)` so that we can use the store in our app.

Finally, in App.vue, we must update to change the items of the left menu. In the template, we must write the following:

```
<template>
  <IonApp>
    <IonSplitPane content-id="main-content">
      <ion-menu content-id="main-content" type="overlay">
        <ion-content>
          <ion-list id="unit-list">
            <ion-list-header>Calculator</ion-list-header>

            <ion-menu-toggle
...
      <ion-router-outlet id="main-content"></ion-router-
        outlet>
    </IonSplitPane>
  </IonApp>
</template>
```

The ion-menu-toggle component contains the menu items we can click or tap on to go to the given page, as specified by the router-link prop. The ion-router-outlet component is where the pages we created earlier are created. The ion-icon component lets us show the icon for each entry.

Next, we will add the imports for App.vue by writing the following code:

```
<script lang="ts">
import {
  IonApp,
  IonContent,
  IonIcon,
  IonItem,
  IonLabel,
  IonList,
  IonListHeader,
  IonMenu,
  IonMenuToggle,
  IonRouterOutlet,
  IonSplitPane,
```

```
} from "@ionic/vue";
import { computed, defineComponent, ref, watch } from
  "vue";
import { RouterLink, useLink, useRoute } from "vue-router";
import { cashOutline, homeOutline } from "ionicons/icons";
...
</script>
```

We will now add the component logic by writing the following code:

```
export default defineComponent({
  name: "App",
  components: {
    IonApp,
    IonContent,
    IonIcon,
    IonItem,
    IonLabel,
    IonList,
    IonListHeader,
    IonMenu,
    IonMenuToggle,
    IonRouterOutlet,
    IonSplitPane,
  },
  setup() {
    const selectedIndex = ref(0);
    const appPages = [
      ...
      {
        title: "Tips Calculator",
        url: "/tips-calculator",
        iosIcon: cashOutline,
        mdIcon: cashOutline,
      },
    ];
```

```
    const route = useRoute();

  return {
    selectedIndex,
    appPages,
    cashOutline,
    route,
  };
},
});
```

Here, we registered the components and added the `appPages` property to render the items. It is not a reactive property as we didn't create it with reactive, but we can use it our template since we returned it. Now, we will add some styles by writing the following code:

```
<style scoped>
...
.selected {
  font-weight: bold;
}
</style>
```

Next, we will add some global styles by writing the following code:

```
<style>
ion-menu-button {
  color: var(--ion-color-primary);
}

#container {
  text-align: center;
  position: absolute;
  left: 0;
  right: 0;
}

#container strong {
  font-size: 20px;
```

```
    line-height: 26px;
}

#container p {
    font-size: 16px;
    line-height: 22px;
    color: #8c8c8c;
    margin: 0;
}

#container a {
    text-decoration: none;
}
</style>
```

By creating the project, we have learned how to use the Composition API, which is used by Ionic to create the Vue project. We also learned how to add type annotations with TypeScript to our JavaScript code to prevent data type errors in our code. And finally, we learned how to create mobile apps from web apps with Ionic.

Summary

With Ionic Vue, we can create mobile apps easily with Vue 3. It makes use of the composition API, TypeScript, and Vue Router, along with the components provided by Ionic, to create good-looking apps that can work as web or mobile apps. It also comes with all the tools required to preview the app in a device or emulator and build it into an app package that we can deploy to app stores.

With the Composition API, we can add the logic like we can with the Vue Options API, but we can add them all with functions instead of referencing them. Ionic Vue also makes TypeScript the default language of components. This lets us prevent type errors at compile time to reduce the chance of type errors happening at runtime. This is a convenience feature that reduces frustration with JavaScript development. We made use of interfaces, union and intersection types, and type aliases to define types for objects.

In the next chapter, we will look at how to build a travel booking app with PrimeVue and Express.

6
Building a Vacation Booking App with the PrimeVue UI Framework

In *Chapter 5, Building a Multipurpose Calculator Mobile App with Ionic*, we built a mobile app with the Ionic mobile app framework, which is built upon Vue.js. However, so far in this book, we have not built any web apps using the UI libraries or frameworks that are based on Vue.js. Additionally, we have not built anything that has its own backend. A backend is definitely something that is required in most systems because we need somewhere to store our data, authenticate users, run background tasks, and more. In this chapter, we will build a vacation booking application with the PrimeVue UI framework.

We will be using the Vue 3 frontend for administration and another frontend for users to add their bookings. We will also include a simple backend to authenticate any administrators before they carry out tasks that can only be done by them. To keep the project as simple as possible, the frontend for the general public won't require authentication.

In this chapter, we will focus on the following topics:

- Building a frontend with the PrimeVue UI framework
- Building a simple backend with Express for authentication
- Persisting data in the backend with SQLite
- Using Vue Router for authentication on the frontend
- Form validation with Vee-Validate and Yup

Technical requirements

The code for this chapter is located at `https://github.com/PacktPublishing/-Vue.js-3-By-Example/tree/master/Chapter06`.

PrimeVue is a UI framework based on Vue 3. This means we can use it for Vue 3 apps. Frameworks that are based on Vue 2 cannot be used by Vue 3 apps because the API has gone through significant changes. The underlying code of Vue 3 is also different from Vue 2. PrimeVue includes a number of common components that we use in web apps, such as text inputs, buttons, menu bars, tables, and more. It is very comprehensive in terms of what is included. Additionally, it comes with styles for items in the form of themes. This means that we can use the built-in components right away. Since PrimeVue is made for Vue 3, we can simply register the components, import the CSS, and use the components in our code. We can also register them either locally or globally depending on which components we need for the given components.

Understanding PrimeVue

PrimeVue comes with styles for the inputs and various kinds of text, such as validation errors and buttons. It also comes with flexbox helpers, which we can use to set the spacing for the components easily. This is very useful because we can simply use the CSS classes that PrimeVue comes with to set the position and spacing of our components.

So far in this book, we have not used any libraries to make form validation more convenient. Form validation is something that we have to do a lot for most web apps.

Understanding Vee-Validate and Yup

Vee-Validate 4 is a form validation library that is compatible with Vue 3. With it, we can add components that will add validation to our forms on the frontend. We will use it with the Yup data validation library to create the form validation schema, which Vee-Validate can use for validation. With the `Form` component, we can add a form that enables us to do form validation with Vee-Validate 4. Then, the `Field` component can be used to validate the form control components that come with PrimeVue. It does this by wrapping around them and passing the slot props that come with the `Field` components as props to the input components.

The Yup library will be used with Vee-Validate to let us validate form values easily without writing all the code from scratch. It lets us create form validation schema objects that we can pass into forms created with Vee-Validate to add form validation.

Understanding Express

To create a simple backend to store data, we use the **Express framework**. This is a very simple Node.js backend framework, which allows us to create a simple backend quickly. To store data, we will use an SQLite database to keep the project simple. We will use the Express framework to create an API to enable the frontend to make HTTP requests to it. We let them make requests by exposing the API endpoint, which the frontend can use by adding routes into it. Each route has a handler function that handles the data submitted by the frontend. We get request data from the HTTP requests made by the frontend, which includes the headers and the body, and we use them in the route handlers to get and store the data the way we want.

Connecting the frontend and the backend

To make the frontend app communicate with the backend app, we'll need to enable cross-domain communication on the backend so that the traffic from the frontend can go through to the backend. This can easily be done with the **Cross-Origin Resource Sharing (CORS)** middleware that we will add to our Express app.

To work with SQLite databases, we use the `sqlite3` library, which lets us manipulate the SQLite database within the Node.js apps. We can make queries and run SQL commands to insert or remove data from our database.

Additionally, we will have simple authentication for the admin frontend. We will check the username and password for the admin login, and if it's valid, we can issue a token and send it to the frontend. Then, the frontend will use the token, which is stored in the header, to check whether the request can be made from the frontend. We add authentication for the *admin-only* routes only, so we only need to check the token for the routes that requires authentication before loading them.

To create and check the token, we use the `jsonwebtoken` library. This allows us to create a token and sign it with a secret string. It also enables us to check the token with a secret to see whether it is valid. We put the `jsonwebtoken` library inside a middleware that is run before the route handler to do the check.

If the token is valid, then we call a function to proceed to the route handler. Otherwise, we send a `401` status back to the client.

Now, we are going to build the project.

Creating the vacation booking project

To create the vacation booking application, we need to create subprojects for the frontend, the admin frontend, and the backend. To create the `frontend` and `admin frontend` project scaffolds, we use the Vue CLI. To create the `backend` folder, we use the `Express Generator` global package.

Follow these steps to set up the project:

1. First, create the `travel-booking-app` folder to house all of the projects.

2. Next, create the `admin-frontend`, `frontend`, and `backend` folders inside the main folder.

3. Go into the `admin-frontend` folder and run the following command:

   ```
   npx vue create
   ```

   This will add the scaffolding code for the Vue project inside the `admin-frontend` folder.

4. If you are asked to create the project in the current folder, select `Y`. Then, when you're asked to choose the Vue version of the project, choose `Vue 3`.

   Likewise, run the Vue CLI in the same way for the `frontend` folder.

5. To create an Express project, run the Express application generator app. To do this, go into the `backend` folder and run the following command:

```
npx express-generator
```

The preceding command will add all of the files required for our project inside the `backend` folder. If you get an error, then try running `express-generator` as an `administrator`.

Now that we have created the project scaffold files and folders, we are ready to start working on the backend.

Creating the backend

Now that we have created the project folders with the scaffolding code, we can start working on the project code. We will start with the backend since we need it for both frontends.

To get started, let's add a few libraries that are needed to manipulate the SQLite database and add authentication to our app. Additionally, we need the library to add CORS to our app.

To install all of them, run the following command:

```
npm i cors jsonwebtoken sqlite3
```

After installing the packages, we are ready to work on the code.

Adding authentication middleware

First, we add our middleware which we will use to check the token. We can do this easily with the `jsonwebtoken` library. This has the `verify` method to check the token.

To add the middleware, create the `middlewares` folder in the backend folder, and then add the `verify-token.js` file. Next, add the following code for the middleware:

```
const jwt = require('jsonwebtoken');

module.exports = (req, res, next) => {
  const token = req.get('x-token')
  try {
    jwt.verify(token, 'secret');
    next()
```

```
  } catch (err) {
    res.status(401)
  }
}
```

Here, we get the x-token request header with the req.get method. Then, we call jwt.verify with the returned token and secret, to verify that the token is valid. Then, we call next if it is valid. If it is not valid, an error will be thrown, and the catch block will be run. res.status with 401 is run to return the 401 response to the frontend since the token isn't valid in this scenario.

The module.exports property is set to the middleware function as the value, which we are exporting. Exporting the function makes it available in the other modules in our backend app.

Add routes to handle requests

Next, we will add the router modules with the routes. First, add the routes to manipulate the bookings. To do this, add the bookings.js file to the routes folder. Inside the file, write the following code:

```javascript
const express = require('express');
const sqlite3 = require('sqlite3').verbose();
const router = express.Router();
const verifyToken = require('../middlewares/verify-token')


router.get('/', (req, res, next) => {
  const db = new sqlite3.Database('./db.sqlite');
  db.serialize(() => {
    db.all(`
      SELECT
        bookings.*,
        catalog_items.name AS catalog_item_name,
        catalog_items.description AS catalog_item_description
      FROM bookings
      INNER JOIN catalog_items ON catalog_items.id =
        bookings.catalog_item_id
```

```
      `,
      [],
      (err, rows = []) => {
        res.json(rows)
      });
    })
  db.close();
});
...
```

Here, we import the required modules including the `verify-token middleware` file that we just created.

The `router.get` method allows us to create a GET request API endpoint. The path is in the first argument. It is the path for the route and it is relative to the path of the router. So, the router's route path is the first segment, and the path in the first argument of `router.get` forms the rest of the URL.

The second argument of the `router.get` method is the route handler. The `req` parameter is an object that has the request data. The `res` parameter is an object that lets us send various kinds of responses to the frontend. We get the database with the `sqlite3.Database` constructor with the path to the database file.

Next, we call the `db.serialize` function so that we can run the code inside the callback in sequence.

The `db.all` method gets all the results returned from the query. The string is the SQL command that retrieves all the data from the bookings table, which we will create with our own SQL code. The `bookings` table is joined with the `catalog_items` table so that the `vacation` package data is associated with the booking.

The second argument of `db.all` is the extra parameters that we want to pass in, which we ignore by passing in an empty array. Then, in the final argument, we have the function with the `err` parameter with the errors in object form. The `rows` parameter has the results from the query. In the callback, we call `res.json` to return the JSON response with the `rows` parameter data.

Then, we call `db.close` to close the database connection once the required operation is done.

Next, we will create a POST route. This will allows us to run an INSERT SQL command to insert an entry into the bookings table. Add the following code:

```
...
router.post('/', (req, res) => {
  const db = new sqlite3.Database('./db.sqlite');
  const { catalogItemId, name, address, startDate, endDate } =
  req.body
  db.serialize(() => {
    const stmt = db.prepare(`
      INSERT INTO bookings (
        catalog_item_id,
        name,
        address,
        start_date,
        end_date
      ) VALUES (?, ?, ?, ?, ?)
    `);
    stmt.run(catalogItemId, name, address, startDate, endDate)
    stmt.finalize();
    res.json({ catalogItemId, name, address, startDate,
      endDate })
  })
  db.close();
});
...
```

Here, we get the body properties with the req.body property. We get all the properties we want to insert into the entry. Next, we create a prepared statement with the INSERT statement. The values at the end are question marks; this means they are placeholders where we can place our own values when we run the stmt.run method.

Prepared statements are useful because they enable us to pass in values to our SQL commands securely. The values are all sanitized so that malicious code cannot run inside the code. We run stmt.run to run the prepared statement with the values we want to replace the placeholder with. We then call stmt.finalize to finalize the operation by writing the data. Next, we call res.json to return the JSON response to the frontend as a response. Then, we call db.close to close the database connection again.

Next, we will create a DELETE endpoint with the `router.delete` method. To do this, write the following code:

```
...
router.delete('/:id', verifyToken, (req, res) => {
  const db = new sqlite3.Database('./db.sqlite');
  const { id } = req.params
  db.serialize(() => {
    const stmt = db.prepare("DELETE FROM bookings WHERE id =
(?)");
    stmt.run(id)
    stmt.finalize();
    res.json({ status: 'success' })
  })
  db.close();
});
...
```

Here, we have the `/:id` path. `:id` is the URL parameter placeholder for the route. We also have the `verifyToken` middleware that we imported at the top of the `booking.js` file. We can use this to verify the token before proceeding to run the code for the route handler. This means that this route is an authenticated route that requires a token in the header for the API endpoint call to succeed.

In the route handler, we get the `id` URL parameter from the `req.params` property. Then, we call `db.serialize`, as we did with the previous route. In the callback, we have the prepared statement, so we can issue a DELETE SQL command with the `id` value that we set in the `stmt.run` method. Then, we call `stmt.finalize`, `res.json`, and `db.close` just as we did in the other routes.

Finally, at the end of the `booking.js` file, let's add the following:

```
module.exports= = router;
```

Adding the preceding statement allows us to import it into another file to register the router. Registering the router will make it accessible.

Next, we will create the `catalog.js` file in the `routes` folder. This file is a `router` module with API endpoints to add our vacation packages. First, we start as follows:

```
const express = require('express');
const router = express.Router();
const sqlite3 = require('sqlite3').verbose();
const verifyToken = require('../middlewares/verify-token')

router.get('/', (req, res,) => {
  const db = new sqlite3.Database('./db.sqlite');
  db.serialize(() => {
    db.all("SELECT * FROM catalog_items", [], (err, rows = [])
      => {
      res.json(rows)
    });
  })
  db.close();
});
...
```

This is almost the same as the GET route in `bookings.js`; however, here, we retrieve all of the items from the `catalog_items` table.

Next, let's add a POST route to add an entry into the `catalog_items` table. Write the following code:

```
...
router.post('/', verifyToken, (req, res,) => {
  const { name, description, imageUrl } = req.body
  const db = new sqlite3.Database('./db.sqlite');
  db.serialize(() => {
    const stmt = db.prepare(`
    INSERT INTO catalog_items (
      name,
      description,
      image_url
    ) VALUES (?, ?, ?)
    `
```

```
  );
    stmt.run(name, description, imageUrl)
    stmt.finalize();
    res.json({ status: 'success' })
  })
  db.close();
});
...
```

Here, we have `verifyToken` in the second argument to check the token in this route before running the route handler in the third argument.

Next, we add a route that enables us to delete a `catalog_items` entry. We do this using the following code:

```
...
router.delete('/:id', verifyToken, (req, res,) => {
  const { id } = req.params
  const db = new sqlite3.Database('./db.sqlite');
  db.serialize(() => {
    const stmt = db.prepare("DELETE FROM catalog_items WHERE
      id = (?)");
stmt.run(id)
stmt.finalize();
res.json({status:'success'})
db.close();
});
...
```

Finally, we export the router:

```
module.exports = router;
```

This module isn't much different from `booking.js`.

Next, we delete the content of the `routes/users.js` file or create it if it doesn't exist. Then, we add the following code:

```
const express = require('express');
const jwt = require('jsonwebtoken');
const router = express.Router();

router.post('/login', (req, res) => {
  const { username, password } = req.body
  if (username === 'admin' && password === 'password') {
    res.json({ token: jwt.sign({ username }, 'secret') })
  }
  res.status(401)
});

module.exports= = router;
```

This is where we check whether the `username` and `password` for the admin user are valid. We only have one user to check here to keep the project simple. We get the `username` and `password` from the `req.body` object, which has the JSON request object. Then, we check for the `username` and `password` with the `if` statement, and if the expression in `if` returns `true`, we call `jwt.sign` to create a token with the token data in the first argument and `secret` in the second argument. Then, we return the response with the authentication token with `res.json`.

Otherwise, we call `res.status` with `401` to return a `401` response, as the `username` or `password` are not valid.

Next, we register our `router` modules and global middleware in `app.js`. To do this, we write the following code:

```
...
const indexRouter = require('./routes/index');
const usersRouter = require('./routes/users');
const catalogRouter = require('./routes/catalog');
const bookingsRouter = require('./routes/bookings');

const app = express();
const cors = require('cors')
...
```

```
app.use('/users', usersRouter);
app.use('/catalog', catalogRouter);
app.use('/bookings', bookingsRouter);
```

We import the `router` modules that we exported earlier with the last line of the `router` files using `require`. Then, we import the `cors` module:

```
const cors = require('cors')
```

We call `app.use` to add the `cors` middleware and then the `router` modules. In the last three lines, we pass in `path` as the first argument and the `router` module as the second argument. This allows us to access the endpoints that we created earlier. With the `cors` module, we can enable cross-domain communication in our Express app.

Next, let's create our SQL script so that we can drop and create the tables easily. To do this, create the `db.sql` file in the `backend` folder and write the following code:

```
DROP TABLE IF EXISTS bookings;
DROP TABLE IF EXISTS catalog_items;

CREATE TABLE catalog_items (
    id INTEGER NOT NULL PRIMARY KEY,
    name TEXT NOT NULL,
    description TEXT NOT NULL,
    image_url TEXT NOT NULL
);

CREATE TABLE bookings (
    id INTEGER NOT NULL PRIMARY KEY,
    catalog_item_id INTEGER NOT NULL,
    name TEXT NOT NULL,
    address TEXT NOT NULL,
    start_date TEXT NOT NULL,
    end_date TEXT NOT NULL,
    FOREIGN KEY (catalog_item_id) REFERENCES catalog_items(id)
);
```

Here, we have created the `bookings` and `catalog_items` tables. Each of these tables has various fields. `TEXT` creates a text column. `NOT NULL` makes the column non-nullable. `PRIMARY KEY` indicates that the column is a primary key column. `FOREIGN KEY` indicates that one column is a foreign key for another column.

We can run the SQL code by installing the DB Browser for the SQLite program, which can be downloaded at `https://sqlitebrowser.org/`, and then creating `db.sqlite` in the `backend` folder. Then, we can go to the **Execute SQL** tab and paste the code into the text input. Following this, we can select all the text and press *F5* to run the code. This will drop any existing `bookings` and `catalog_items` tables and create them again. For changes for the database to be written to disk, you have to save them. To do this, click on the **File** menu and then click on **Write Changes**. We can also press the *Ctrl + S* keyboard combination to save the changes.

Finally, to make our app run and restart automatically when we change the code, we can install the `nodemon` package globally. To do this, run the following command:

```
npm i -g nodemon
```

Then, in the `package.json` file, change the `script.start` property's value to the following code:

```
{
  ...
  "scripts": {
    "start": "nodemon ./bin/www"
  },
  ...
}
```

We can run `npm start` with `nodemon` instead of the regular node executable, which means the app will restart automatically when we change any code file and save it.

Now that we have created a basic backend for the frontends to consume, we can move on to create our frontend apps with PrimeVue and Vue 3.

Creating the admin frontend

Now that the backend app is complete, we can move on to work on the admin frontend. We already created the Vue 3 project for the admin frontend in the `admin-frontend` folder earlier, so we just need to install packages that we require and work on the code. We will need the PrimeVue packages – that is, the Vee-Validate, Vue Router, Axios, and Yup packages.

To install them, run the following command in the `admin-frontend` folder:

```
npm i axios primeflex primeicons primevue@^3.1.1 vee-validate@
next vue-router@4 yup
```

Axios allows us to make HTTP requests to the backend. Vue Router lets us map URLs to the `page` components. Vee-Validate and Yup allow us to easily add form validation to our forms, and the remaining packages are the PrimeVue packages.

Creating the admin frontend pages

After installing the packages, we can start working on the code. First, we will work on the components. In the `components` folders, add the `CatalogForm.vue` file and write the following code:

```
<template>
  <Form @submit="onSubmit" :validation-schema="schema">
    <Field v-slot="{ field, errors }" v-model="name"
      name="name">
      <div class="p-col-12">
        <div class="p-inputgroup">
          <InputText
            placeholder="Name"
            :class="{ 'p-invalid': errors.length > 0 }"
            v-bind="field"
          />
        </div>
        <small class="p-error" v-if="errors.length > 0">
          Name is invalid.
        </small>
      </div>
    </Field>
```

```
    ...
  </Form>
</template>
```

Here, we have the Form component from the Vee-Validate package to add a form with form validation. The submit event is only emitted when all of the form values are valid. We will register the Form component later. The validation-schema prop is set to the validation schema object created by Yup.

Inside the Form component, we have a Field component, which is also provided by the Vee-Validate package. We will also register this component globally later so that we can use it. Inside the Field component, we have the InputText component to add an input field into our app. To enable form validation for the InputText component, we pass in the field object to the slot props and pass the whole thing as the value of the v-bind directive. The v-bind directive allows Vee-Validate to handle the form values and add validation to our form field. The errors array gives us any validation errors that might have occurred.

The p-col-12 class is provided by PrimeVue's PrimeFlex package. It lets us set the width of a div tag to full width, which means it takes 12 columns out of 12 on the page. With the p-inputgroup class, we can create an input group. The p-error class styles the text color to red so that we can show form validation messages in a way that is easy for the user to see. The p-invalid class makes the edge of the input red. We only change it to red if the error's length is bigger than 0 since this means there are validation errors, and we only show the smaller element when the error's length is bigger than 0.

The Field component has a v-model directive to bind the inputted value to the corresponding reactive properties. We also have a name attribute that is also used as the property name of the value parameter for the submit event handler, which has the inputted values. These values are always valid since the submit handler is only run when all of the form values are valid.

With the name field, we can enter the name of the vacation package.

Next, we need to add a text area to allow users to enter a description for the vacation package. To do this, write the following code:

```
<template>
  <Form @submit="onSubmit" :validation-schema="schema">
    ...
    <Field v-slot="{ field, errors }" v-model="description"
      name="description">
```

```
        <div class="p-col-12">
          <div class="p-inputgroup">
            <Textarea
              placeholder="Description"
              :class="{ 'p-invalid': errors.length > 0 }"
              v-bind="field"
            />
          </div>
          <small class="p-error" v-if="errors.length > 0">
            Description is invalid
          </small>
        </div>
      </Field>
      ...
    </Form>
</template>
```

This is almost the same as the name field; however, in this scenario, we switch out the InputText component for the Textarea component. We also change the v-model and name values. The Textarea component is from the PrimeVue package, which renders into a textarea element with its own styles.

Next, we add the image URL field so that we can add an image URL for the vacation package. We just let the user enter the image URL to make our project simpler. To add the field to the Form component, write the following code:

```
<template>
  <Form @submit="onSubmit" :validation-schema="schema">
    ...
    <Field v-slot="{ field, errors }" v-model="imageUrl"
      name="imageUrl">
      <div class="p-col-12">
        <div class="p-inputgroup">
          <InputText
            placeholder="Image URL"
            :class="{ 'p-invalid': errors.length > 0 }"
            v-bind="field"
          />
```

```
        </div>
        <small class="p-error" v-if="errors.length > 0">
           Image URL is invalid.
        </small>
      </div>
    </Field>
    . . .
  </Form>
</template>
```

This is just another text input with a different name and `v-model` value. Finally, let's add a `submit` button to the form using the following code:

```
<template>
  <Form @submit="onSubmit" :validation-schema="schema">
    . . .
    <div class="p-col-12">
      <Button label="Add" type="submit" />
    </div>
  </Form>
</template>
```

The `Button` component is from the PrimeVue package, which we will register globally later to make it available everywhere.

Next, we add the `component` options object. We use the `component` options API to create our components. First, we import everything and create the form validation schema with the Yup library. To add the code, write the following in `components/CatalogForm.vue`:

```
<script>
import * as yup from "yup";
import axios from "axios";
import { APIURL } from "@/constants";

const schema = yup.object().shape({
  name: yup.string().required(),
  description: yup.string().required(),
  imageUrl: yup.string().url().required(),
```

```
});

export default {
  name: "BookingForm",
  data() {
    return {
      name: "",
      description: "",
      imageUrl: "",
      schema,
    };
  },
  ...
};
</script>
```

Here, we create the schema object with the `yup.object` method, which allows us to validate an object with some properties. The `validation` schema is separate from the `v-model` binding. The property of the object that we pass into the `shape` method has to match the `name` attribute's value of the `Field` component.

To validate the value of the field with `name` set to `name`, we set the `name` property to `yup.string().required()` to ensure that the `name` field is a string and has a value. We set the same value for `description`. The `imageUrl` value is set to `yup.string().url().required()` to ensure that the inputted value is a URL and that it is filled in.

The `data` method returns the schema so that we can use the `validation-schema` prop of the `Form` component.

To finish the component, we add the `onSubmit` method, which is called when the `submit` event is emitted by the `Form` component:

```
<script>
...
export default {
  ...
  methods: {
    async onSubmit(value) {
      const { name, description, imageUrl } = value;
```

```
        await axios.post(`${APIURL}/catalog`, {
          name,
          description,
          imageUrl,
        });
        this.$emit("catalog-form-close");
      },
    },
  };
</script>
```

Here, we simply take the `property` values from the `value` parameter, which has the valid form field values. Then, we make a POST request to the catalog endpoint with the JSON payload passed into the second argument. Following this, we call the `this.$emit` method to emit the `catalog-form-close` event to signal to the dialog component that this form will be housed in to close.

Add a top bar and menu bar

Next, we will add a `top bar` component into our app. To do this, create `TopBar.vue` in the `src/components` folder. Then, add the following template code into the file:

```
<template>
  <Menubar :model="items">
    <template #start>
      <b>Admin Frontend</b>
    </template>
  </Menubar>
</template>
```

The `Menubar` component is provided by the PrimeVue component. We can use it to add a menu bar with some items that we can click on to navigate to different pages. The `model` prop is set to the `items` reactive property, which is an array of menu item objects that we will add shortly. The `start` slot lets us add items to the left-hand side of the menu bar. We can put some bold text into the slot and it'll be displayed on the left-hand side.

Next, we can add a `component` object for the component. To add it, write the following code:

```
<script>
export default {
  name: "TopBar",
  props: {
    title: String,
  },
  data() {
    return {
      items: [
        {
          label: "Manage Bookings",
          command: () => {
            this.$router.push("/bookings");
          },
        },
...
  methods: {
    logOut() {
      localStorage.clear();
      this.$router.push("/");
    },
  },
  beforeMount() {
    document.title= = this.title;
  },
};
</script>
```

Here, we register the `title` prop, which we use to set the `document.title` value. The `document.title` property sets the title on the top bar. In the `data` method, we return an object with the item's reactive property. This is set to an object with the `label` and `command` properties. The `label` property is shown in the menu bar item for the user. The item is shown as a link. The `command` method is run when we click on the item.

With the `this.$router.push` method, we can navigate to the page that is mapped to the given URL. The `logOut` method navigates back to the page mapped to the / path, which is the *login* page that we will discuss later. Additionally, we clear the local storage so that we can clear the authentication token.

In the `beforeMount` hook, we set the `document.title` property to the value of the `title` prop.

Add shared code to deal with requests

Next, let's write the code for the Axios request interceptor to let us add the authentication token to the `x-token` request header of all requests aside from when we make requests to the `/login` endpoint. To do this, create the `src/plugins` folder and add `axios.js` to it. Then, inside this file, write the following code:

```
import axios from 'axios'
import { APIURL } from '@/constants'

axios.interceptors.request.use((config) => {
  if (config.url.includes(APIURL)
    && !config.url.includes('login')) {
    config.headers['x-token'] = localStorage.getItem('token')
    return config
  }
  return config
}, (error) => {
  return Promise.reject(error)
})
```

Here, we check the URL that the request is being made to by retrieving the URL with the `config.url` property. Then, if we make any requests to an endpoint other than `/login`, we set the `x-token` request header:

```
config.headers['x-token'] = localStorage.getItem('token')
```

Note that we get the token from local storage and set it to the value of `config.headers['x-token']`. The `config.headers` property is an object with the request headers. The second argument is the request error handler. Here, we simply return a rejected promise with `Promise.reject` so that we can handle the error.

Next, we add Vue Router routes to our routes. We stay in the `src/plugins` folder and create a `vue-router.js` file. Then, we add the following code to the file:

```
import { createWebHashHistory, createRouter } from 'vue-
  router'
import Login from '../views/Login.vue'
import Bookings from '../views/Bookings.vue'
import Catalog from '../views/Catalog.vue'

const beforeEnter = (to, from, next) => {
  try {
    const token = localStorage.getItem('token')
    if (to.fullPath !== '/' && !token) {
      return next({ fullPath: '/' })
    }
    return next()
  } catch (error) {
    return next({ fullPath: '/' })
  }
}

const routes = [
  { path: '/', component: Login },
  { path: '/bookings', component: Bookings, beforeEnter },
  { path: '/catalog', component: Catalog, beforeEnter },
]

const router = createRouter({
  history: createWebHashHistory(),
  routes,
})
export default router
```

We add the `beforeEnter` function to check for the token if we go to any page on the frontend other than the home page. We can check the path that the user tries to go to with the `to.fullPath` property. If it is anything other than `'/'` and there is no token in local storage, then we call `next` with an object, and the `fullPath` property set to `'/'` to go to the login page. Otherwise, we call `next` with no argument to go to the page we are supposed to go to. If we have an error, then we also go to the login page, as you can see from the code in the `catch` block.

Next, we have the `routes` array with the route definitions. This has the `route` path in the `path` property, and `component` is the component that the path maps to. The `beforeEnter` property is added to the last two route objects so that we can only go there once we are logged in.

Then, to create the `router` object, we call `createRouter` with an object with the `history` property set to the object returned by the `createWebHashHistory` function; this is so that we can keep the hash between the hostname and the rest of the URL. We set the `routes` property to the `routes` array in order to register the routes. This is so we can see the right component when we go to the routes.

Finally, we export the `router` object as a default export so that we can add the router object to our app later with `app.use`.

Next, we create the `views` folder inside the `src` folder. This means we can add the pages that users can go to. Now, let's add a page to allow the admins to manage any bookings by adding the `Bookings.vue` file to the `src/views` folder. We open the file and add the following template to the component. This is so that we can add the `TopBar` component that we created earlier:

```
<template>
  <TopBar title="Manage Bookings" />
  <div class="p-col-12">
    <h1>Manage Bookings</h1>
  </div>
  <div class="p-col-12">
    <Card v-for="b of bookings" :key="b.id">
      <template #title>{{ b.name }} </template>
      <template #content>
        <p>Address: {{ b.address }}</p>
        <p>Description: {{ b.description }}</p>
        <p>Start Date: {{ b.start_date }}</p>
```

```
      <p>End Date: {{ b.end_date }}</p>
    </template>
    <template #footer>
      <Button
        icon="pi pi-times"
        label="Cancel"
        class="p-button-secondary"
        @click="deleteBooking(b.id)"
      />
    </template>
  </Card>
</div>
</template>
```

Note that we add the heading for the page using the h1 element. Then, we add the Card component to display the bookings to the admin. The Card component is provided by PrimeVue, and we will register it later. We use the v-for directive to render the bookings array into multiple Card components. The key prop is set to a unique ID so that Vue 3 can distinguish each item properly.

We populate the title, content, and footer slots with different content. The footer slot has a Button component that runs the deleteBooking function when we click on the button. The icon prop allows us to set the icon on the left-hand side of the button. The label prop has the button text on the right-hand side of the icon. With the p-button-secondary class, we can set the color of the button.

Next, we can add the component options object with the getBooking and deleteBooking methods to retrieve bookings and delete bookings via the backend API, respectively. To add them, write the following code:

```
<script>
import axios from "axios";
import { APIURL } from "@/constants";
import TopBar from "@/components/TopBar";

export default {
  name: "Bookings",
  components: {
```

```
      TopBar,
    },
  data() {
    return {
      bookings: [],
    };
  },
  ...
  beforeMount() {
    this.getBookings();
  },
};
</script>
```

We also register the `TopBar` component in the `components` property. The `getBookings` method calls `axios.get` to make a GET request and sets the value of the `this.bookings` reactive property to the response object.

`bookings` is stored inside the `data` property of the object that is returned as the resolved value of the returned promise.

Likewise, we call `axios.delete` inside the `deleteBooking` method to make a DELETE request to delete the items. Then, we call `this.getBookings` to get the data again. We also call `this.getBookings` in the `beforeMount` hook to get the data when the page loads.

Next, we add a page to allow admins to manage the vacation package items. To do this, let's add the `Catalog.vue` file to the `src/views` folder. Then, inside the file, write the following:

```
<template>
  <TopBar title="Manage Vacation Packages" />
  <div class="p-col-12">
    <h1>Manage Vacation Packages</h1>
  </div>
  <div class="p-col-12">
    <Button label="Add Vacation Package"
      @click="displayCatalog= = true" />
    <Dialog header="Add Vacation Package" v-
```

```
        model:visible="displayCatalog">
        <CatalogForm
          @catalog-form-close="
            displayCatalog= = false;
            getCatalog();
          "
        />
      </Dialog>
    </div>
    ...
</template>
```

Here, we add the `TopBar` component to display the top bar; `h1` displays a heading. Next, we add a button that will let us show the dialog by setting `displayCatalog` to `true`. Then, we display the `Dialog` component by setting the `v-model` directive with the visible modifier to the `displayCatalog` value. Using this, we can control when the `Dialog` component is displayed. The `Dialog` component displays a dialog box, and this component is provided by PrimeVue.

The `header` prop sets the header text for the dialog box. We use `CatalogForm` as the content, and we listen to the `catalog-form-close` event emitted by the `CatalogForm` component. When it is emitted, we set `displayCatalog` to `false` and call `getCatalog` to get the data again:

```
<template>
  ...
  <div class="p-col-12">
    <Card v-for="c of catalog" :key="c.id">
      <template #header>
        <img :alt="c.description" :src="c.image_url" />
      </template>
      <template #title> {{ c.name }} </template>
      <template #content>
        {{ c.description }}
      </template>
      <template #footer>
        <Button
```

```
                icon="pi pi-times"
                label="Delete"
                class="p-button-secondary"
                @click="deleteCatalogItem(c.id)"
          />
        </template>
      </Card>
    </div>
</template>
```

Next, we add the `Card` components that are rendered from the catalog reactive property with the `v-for` directive to render the catalog entries. The remaining code is similar to what we had in the `Bookings.vue` file, but now the render properties are different, and `Button` calls a different method when we click on it.

Following this, we add the component object by adding the following code to `src/views/Catalog.vue`:

```
<script>
import axios from "axios";
import { APIURL } from "@/constants";
import TopBar from "@/components/TopBar";
import CatalogForm from "@/components/CatalogForm";
...
  methods: {
    async getCatalog() {
      const{ { data } = await axios.get(`${APIURL}/catalog`);
      this.catalog = data;
    async deleteCatalogItem(id) {
      await axios.delete(`${APIURL}/catalog/${id}`);
      this.getCatalog();
    },
  },
  beforeMount() {
    this.getCatalog();
  },
};
</script>
```

Here, the code is similar to what we had in `src/views/Bookings.vue` except that, here, we make requests to the catalog endpoints to get and delete the data.

Then, we create the last page in the admin frontend app, which is the login page. To add the login page, we add the `Login.vue` file to the `src/views` folder. Then, inside the file, we add the `form` and the `username` field using the following code:

```
<template>
  <Form @submit="onSubmit" :validation-schema="schema">
    <div class="p-col-12">
      <h1>Admin Log In</h1>
    </div>

    <Field v-slot="{ field, errors }" v-model="username"
      name="username">
      <div class="p-col-12">
        <div class="p-inputgroup">
          <InputText
            placeholder="Username"
            :class="{ 'p-invalid': errors.length > 0 }"
            v-bind="field"
          />
        </div>
        <small class="p-error" v-if="errors.length > 0">
          Username is invalid.
        </small>
      </div>
    </Field>
    ...
  </Form>
</template>
```

The `username` field is similar to all of the other fields that we have added before. Next, we add the `password` input and button using the following code:

```
<template>
  <Form @submit="onSubmit" :validation-schema="schema">
    ...
    <Field v-slot="{ field, errors }" v-model="password"
      name="password">
      <div class="p-col-12">
        <div class="p-inputgroup">
          <InputText
            placeholder="Password"
            type="password"
            :class="{ 'p-invalid': errors.length > 0 }"
            v-bind="field"
          />
        </div>
        <small class="p-error" v-if="errors.length > 0">
          Password is invalid
        </small>
      </div>
    </Field>

    <div class="p-col-12">
      <Button label="Log In" type="submit" />
    </div>
  </Form>
</template>
```

We set the `type` prop to `password` to make the field a password input. The button's `type` prop is set to `submit` so that we can trigger the `submit` event when we click on it and all of the form values remain valid.

Next, we add the component object portion of the `Login.vue` file, which has the `onSubmit` method to make the login request:

```
<script>
import * as yup from "yup";
import axios from "axios";
```

```
import { APIURL } from "@/constants";

const schema = yup.object().shape({
  username: yup.string().required(),
  password: yup.string().required(),
});

export default {
  name: "Login",
  data() {
    return {
      username: "",
      password: "",
      schema,
    };
  },
  methods: {
    async onSubmit(values) {
      const { username, password } = values;
      try {
        const {
          data: { token },
        } = await axios.post(`${APIURL}/users/login`, {
          username,
          password,
        });
        localStorage.setItem("token", token);
        this.$router.push("/bookings");
      } catch (error) {
        alert("Login failed");
      }
    },
  },
};
</script>
```

We create the schema object with the validation schema, which is similar to the other schemas we have used previously. Then, we add that to the object we returned in the data method. The onSubmit method takes the username and password properties from the value parameter so that we can use it to make the POST request to the /users/ login endpoint.

Once we have done that, we get a token from the response if the request is successful along with the localStorage.setItem method. Next, we call the this.$router. push method to redirect to the /bookings URL. If there are any errors, we show an alert with the "Login failed" message.

Next, we add the router-view component provided by Vue Router to App.vue. This is so we can show the page that we created in the routes object. To add it, write the admin frontend:shared code, adding to deal with requests" following code:

```
<template>
  <router-view></router-view>
</template>

<script>
export default {
  name: "App",
};
</script>

<style>
body {
  background-color: #ffffff;
  font-family: -apple-system, BlinkMacSystemFont, Segoe UI,
    Roboto, Helvetica,
    Arial, sans-serif, Apple Color Emoji, Segoe UI Emoji,
      Segoe UI Symbol;
  font-weight: normal;
  color: #495057;
  -webkit-font-smoothing: antialiased;
  -moz-osx-font-smoothing: grayscale;
  margin: 0px;
}
</style>
```

We also have a `style` tag to set the font family and set the margin of the body to `0px`, so there is no white space between the elements and the edge of the page.

Next, we add `constants.js` to the `src` folder and then add `APIURL` to it:

```
export const APIURL = 'http://localhost:3000'
```

In `main.js`, we register all of the global components along with the router object we created earlier. We also import the global styles provided by PrimeVue, so everything looks good:

```
import { createApp } from 'vue'
import App from './App.vue'
import PrimeVue from 'primevue/config';
import InputText from "primevue/inputtext";
import Button from "primevue/button";
import Card from 'primevue/card';
import Toolbar from 'primevue/toolbar';
import router from './plugins/vue-router'
import Textarea from 'primevue/textarea';
import Dialog from 'primevue/dialog';
import Menubar from 'primevue/menubar';
import { Form, Field } from "vee-validate";
import "primeflex/primeflex.css";
import 'primevue/resources/themes/bootstrap4-light-blue/theme.css'
import "primevue/resources/primevue.min.css";
import "primeicons/primeicons.css";
import './plugins/axios'
...
app.component("Form", Form);
app.component("Field", Field);
app.use(PrimeVue);
app.use(router)
app.mount('#app')
```

In `package.json`, we change the port that the development server runs on by changing the `script.serve` property to the following:

```
{
  ...
  "scripts": {
    "serve": "vue-cli-service serve --port 8082",
    "build": "vue-cli-service build",
    "lint": "vue-cli-service lint"
  },
  ...
}
```

Now, when we run `npm run serve`, we get the following screenshot:

Figure 6.1 – Admin Frontend

Now that we have created the admin frontend app, all we have left to add is the user frontend.

Creating the user frontend

Now that we have finished with the admin frontend, we will complete this chapter's project by creating the user's frontend. The user frontend is similar to the admin frontend; however, in this case, there is no authentication required to use it.

We will start by installing the same packages that we installed for the admin frontend. Navigate to the frontend folder and run the following command:

```
npm i axios primeflex primeicons primevue@^3.1.1 vee-validate@
next vue-router@4 yup
```

Next, create the `src/components` folder, if it doesn't exist. Then, create the `BookingForm.vue` file inside of `src/components` so that users can add their bookings.

Add the `form` and `name` fields to allow users to enter their name:

```
<template>
  <Form @submit="onSubmit" :validation-schema="schema">
    <Field v-slot="{ field, errors }" v-model="name"
      name="name">
      <div class="p-col-12">
        <div class="p-inputgroup">
          <InputText
            placeholder="Name"
            :class="{ 'p-invalid': errors.length > 0 }"
            v-bind="field"
          />
        </div>
        <small class="p-error" v-if="errors.length > 0">Name
          is invalid </small>
      </div>
    </Field>
    ...
  </Form>
</template>
```

This is very similar to the other text input fields that we added earlier. Then, add the `address` field using the following code:

```
<template>
  <Form @submit="onSubmit" :validation-schema="schema">
    ...
    <Field v-slot="{ field, errors }" v-model="address"
      name="address">
      <div class="p-col-12">
        <div class="p-inputgroup">
          <InputText
            placeholder="Address"
            :class="{ 'p-invalid': errors.length > 0 }"
            v-bind="field"
          />
        </div>
        <small class="p-error" v-if="errors.length > 0"
          >Address is invalid</small
        >
      </div>
    </Field>
    ...
  </Form>
</template>
```

Now, let's add the `Calendar` component that is provided by PrimeVue, which we have not used before in this project. The `Calendar` component lets users pick a date. We can add the `Start Date` field to allow the user to pick the start date of their vacation:

```
<template>
  <Form @submit="onSubmit" :validation-schema="schema">
    ...
    <Field v-slot="{ field, errors }" v-model="startDate"
      name="startDate">
      <div class="p-col-12">
        <label>Start Date</label>
        <div class="p-inputgroup">
          <Calendar
```

```
          inline
          placeholder="Start Date"
          :class="{ 'p-invalid': errors.length > 0 }"
          :minDate="new Date()"
          v-bind="field"
          v-model="startDate"
        />
      </div>
      <small class="p-error" v-if="errors.length > 0">
        Start date is invalid
      </small>
    </div>
  </Field>

  ...

 </Form>
</template>
```

Here, we have the `minDate` prop, which sets the earliest date that the user can pick.
The `inline` prop will make the date picker display on the form instead of in a popup.
Likewise, we can add the End Date field using the following code:

```
<template>
  <Form @submit="onSubmit" :validation-schema="schema">
    ...
    <Field v-slot="{ field, errors }" v-model="endDate"
        name="endDate">
      <div class="p-col-12">
        <label>End Date</label>
        <div class="p-inputgroup">
          <Calendar
            inline
            placeholder="End Date"
            :class="{ 'p-invalid': errors.length > 0 }"
            v-bind="field"
            v-model="endDate"
            :minDate="new Date(+startDate + 24 * 3600 * 1000)"
          />
```

```
      </div>
      <small class="p-error" v-if="errors.length > 0"
        >End date is invalid</small
      >
    </div>
  </Field>
  ...
  </Form>
</template>
```

Here, we set the `minDate` prop to a day after `startDate`. `24 * 3600 * 1000` milliseconds is equivalent to one day. Finally, we add the `submit` button just as we did in our other forms:

```
<template>
  <Form @submit="onSubmit" :validation-schema="schema">
    ...
    <div class="p-col-12">
      <Button label="Book" type="submit" />
    </div>
    ...
  </Form>
</template>
```

Next, we create `schema` by writing the following:

```
<script>
import { Form, Field } from "vee-validate";
import * as yup from "yup";
import axios from "axios";
import { APIURL } from "@/constants";

const schema = yup.object().shape({
  name: yup.string().required(),
  address: yup.string().required(),
  startDate: yup.date().required().min(new Date()),
  endDate: yup
    .date()
```

```
        .required()
        .when(
          "startDate",
          (startDate, schema) => startDate &&
            schema.min(startDate)
        ),
    });

    ...
</script>
```

To validate endDate, we call the when method with the field name that we want to check against. Then, we call schema.min to make sure that the endDate is later than the startDate.

Next, we add the component object to register the selectedCatalogId prop and add the onSubmit method. We write the following code:

```
<script>
...
export default {
  name: "BookingForm",
  components: {
    Form,
    Field,
  },
  props: {
...
  methods: {
    async onSubmit(values) {
      const { name, address, startDate, endDate } = values;
      await axios.post(`${APIURL}/bookings`, {
        name,
        address,
        startDate,
        endDate,
        catalogItemId: this.selectedCatalogId,
      });
```

```
        this.$emit("booking-form-close");
      },
    },
  };
</script>
```

The onSubmit method gets the form field values from the values parameter
and makes a POST request to the bookings endpoint to add a booking. We use
selectedCatalogId to add the booking. Then, we emit the booking-form-close
event to emit an event to the parent to signal the form to close.

Next, we add Vue Router to our app by adding vue-router.js to the
src/plugins folder:

```
import { createWebHashHistory, createRouter } from 'vue-router'
import Catalog from '../views/Catalog.vue'

const routes = [
  { path:'/', component: Catalog },
]

const router = createRouter({
  history: createWebHashHistory(),
  routes,
})

export default router
```

This is very similar to what we had in the admin frontend.

Next, we create a page to show all of the vacation packages to the user by adding the
src/views/Catalog.vue file and then adding the following template code:

```
<template>
  <Card v-for="c of catalog" :key="c.id">
    <template #header>
      <img :alt="c.description" :src="c.image_url" />
    </template>
    <template #title> {{ c.name }} </template>
    <template #content>
```

```
          {{ c.description }}
      </template>
      <template #footer>
        <Button
          icon="pi pi-check"
          label="Book"
          class="p-button-secondary"
          @click="book(c.id)"
        />
...
            :selectedCatalogId="selectedCatalogId"
          />
      </Dialog>
    </template>
  </Card>
</template>
```

Here, we simply render a form from the catalog array. We have a `Dialog` component with the `BookingForm` component inside. We listen to the `booking-form-close` event emitted by it to close the `Dialog` component by setting `displayBookingForm` to `false` and calling `displayMessage` to display the alert. We pass in `selectedCatalogId` as the value of the prop with the same name.

The remainder of the template code is almost the same as what we had previously, except for the property names displayed and the addition of the image inside the header slot.

Next, we add the `component` options object to the same file by writing the following code:

```
<script>
import axios from "axios";
import { APIURL } from "@/constants";
import BookingForm from "../components/BookingForm.vue";

export default {
  name: "Catalog",
  components: {
    BookingForm,
  },
  data() {
```

```
    return {
      selectedCatalogId: undefined,
      displayBookingForm: false,
...
    displayMessage() {
      alert("Booking successful");
    },
  },
  beforeMount() {
    this.getCatalog();
  },
};
</script>
```

We register the `BookingForm` component within the `components` property. The `getCatalog` function gets the vacation package catalog items from the API. The `booking` function sets `displayBookingForm` to `true` to open the `Dialog` component, and `selectedCatalogId` is also set there. The `beforeMount` hook calls `getCatalog` to retrieve the catalog data.

Adding the router view and entry point code

In `App.vue`, we write the following code to add `router-view` and set the same styles that we did in the admin frontend:

```
<template>
  <router-view></router-view>
</template>

<script>
export default {
  name: "App",
};
</script>

<style>
body {
  background-color: #ffffff;
```

```
    font-family: -apple-system, BlinkMacSystemFont, Segoe UI,
      Roboto, Helvetica, Arial, sans-serif, Apple Color Emoji,
      Segoe UI Emoji, Segoe UI Symbol;
    font-weight: normal;
    color: #495057;
    -webkit-font-smoothing: antialiased;
    -moz-osx-font-smoothing: grayscale;
    margin: 0px;
}
</style>
```

Then, we create `constants.js` in the `src` folder and add the following line to add `APIURL`:

```
export const APIURL = 'http://localhost:3000'
```

Then, in `main.js`, we replace the contents of the file with the following code to register the components and the router globally. We also import the styles provided by PrimeVue to make our app look good:

```
import { createApp } from 'vue'
import App from './App.vue'
import PrimeVue from 'primevue/config';
import InputText from "primevue/inputtext";
import Button from "primevue/button";
import Card from 'primevue/card';
import Toolbar from 'primevue/toolbar';
import Calendar from 'primevue/calendar';
import Dialog from 'primevue/dialog';
import router from './plugins/vue-router'
import "primeflex/primeflex.css";
import 'primevue/resources/themes/bootstrap4-light-blue/theme.css'
import "primevue/resources/primevue.min.css";
import "primeicons/primeicons.css";


const app = createApp(App)
app.component("InputText", InputText);
```

```
app.component("Button", Button);
app.component("Card", Card);
app.component("Toolbar", Toolbar);
app.component("Calendar", Calendar);
app.component("Dialog", Dialog);
app.use(PrimeVue);
app.use(router)
app.mount('#app')
```

In `package.json`, we change the port that the development server runs on by changing the `script.serve` property to the following:

```
{
    ...
    "scripts": {
        "serve": "vue-cli-service serve --port 8082",
        "build": "vue-cli-service build",
        "lint": "vue-cli-service lint"
    },
    ...
}
```

Now, when we run `npm run serve`, we get the following screenshot:

Figure 6.2 – The user frontend

With the user frontend created, the vacation booking system is now complete.

Summary

In this chapter, we learned how to use PrimeVue effectively to build a vacation booking application. With PrimeVue, we can create good-looking Vue 3 web apps easily. PrimeVue comes with many useful components that we can add to create our web apps, such as inputs, text areas, tables, dialog boxes, date pickers, and more. It also comes with styles built-in, so we don't have to add any styles from scratch ourselves. Additionally, we can add the PrimeFlex package that is also provided by PrimeVue; with flexbox, we can change the spacing and positions of elements and components easily.

Vee-Validate and Yup allow us to add form validation into our Vue 3 app. This integrates easily with the input components provided by PrimeVue. These two libraries make a lot of the form validation work easy, as we don't have to write all the form validation code ourselves.

To make a simple backend, we used Express to create a simple API to interact with the frontend. We also used the `sqlite3` package to manipulate data with the SQLite databases in our API. Express comes with many add-ons that we can use to add a lot of functionality, such as cross-domain communication. We can also easily add JSON Web Token authentication to our Express app via the `jsonwebtoken` library.

In the next chapter, we will learn how to build a storefront with GraphQL and Vue 3.

7
Creating a Shopping Cart System with GraphQL

In the previous chapter, we built a travel booking system with Vue 3 and Express. This was the first project where we built from scratch our own backend that is used by the frontend. Having our own backend lets us do a lot more stuff that we can't do otherwise—for example, we can save the data that we like in the database that we created ourselves. Also, we added our own authentication system to authenticate the admin user. On the admin frontend, we protect our routes with the `beforeEnter` route guard, which checks for the authentication token before the admin user can log in.

In this chapter, we will take a look at the following topics:

- Introducing the GraphQL **application programming interface (API)**
- Creating a GraphQL API with Express
- Creating the admin frontend
- Creating the customer frontend

Technical requirements

The code for this chapter's project can be found at `https://github.com/PacktPublishing/-Vue.js-3-By-Example/tree/master/Chapter07`.

Introducing the GraphQL API

In the last chapter, we created a backend with Express. The endpoint accepts JSON data as input and returns JSON data as a response. However, it can take any JSON data, which the backend may not expect. Also, there is no easy way to test our API endpoints without the frontend. This is something that we can solve with GraphQL APIs. **GraphQL** is a special query language that makes communication easier between the client and server. GraphQL APIs have a built-in data structure validation. Each property has a data type, which can be a simple or complex type, consisting of many properties with simple data types.

We can also test GraphQL APIs with GraphiQL, which is a web page that lets us make our own GraphQL API requests easily. Since there is a data type validation for each request, it can provide an autocomplete feature, according to the definition of the GraphQL API schema. The schema provides us with all the data type definitions that are used with queries and mutations. Queries are requests that let us query for data with our GraphQL API, while mutations are GraphQL requests that let us change data in some way.

We define queries and mutations explicitly with a schema string. The queries and mutations take input types as data types for the input data, and return data with the specified output data types. Therefore, we will never be in any doubt about the structure of the data that we have to send to make a GraphQL request and will never have to guess as to what kind of data a request will return.

GraphQL API requests are mostly just regular **HyperText Transfer Protocol** (**HTTP**) requests, except that they have a special structure. All requests go to the `/graphql` endpoint by default, and we send queries or mutations as a string value of the `query` property in the JSON requests. The variable values are sent with the `variable` parameter.

The queries and mutations are named, and all the queries and mutations are sent to the resolver functions with the same names in the code, instead of to route handlers. The functions then take the arguments specified by the schema, after which we can get the request data and do what we want with it in our resolver function code.

With Vue 3 apps, we can use specialized GraphQL API clients to make GraphQL API request creation easier. All we have to do to make a request is pass in a string for the queries and mutations, along with the variables that go with the queries and mutations.

In this chapter, we will create a shopping cart system with an admin frontend and a customer frontend with Vue 3. We will then create a backend with Express and the `express-graphql` library that takes GraphQL API requests and stores data in a SQLite database.

Setting up the shopping cart system project

To create the vacation booking project, we had to create subprojects for the frontend, the admin frontend, and the backend. To create the frontend and admin frontend projects, we will use Vue CLI. To create the backend project, we will use the `express-generator` global package.

To set up this chapter's project, we execute the following steps:

1. First, we create a folder to house all the projects, and name it `shopping-cart`.

2. We then create `admin-frontend`, `frontend`, and `backend` folders inside the main folder.

3. Next, we go into the `admin-frontend` folder and run `npx vue create` to add the scaffolding code for the Vue project to the `admin-frontend` folder.

4. If we are asked to create the project in the current folder, we select `Y`, and then when we're asked to choose the Vue version of the project, we choose `Vue 3`. Likewise, we run Vue CLI the same way with the `frontend` folder.

5. To create the Express project, we run the Express application generator app. To run it, we go into the `backend` folder and then run `npx express-generator`.

   This command will add all the files that are required for our project to the `backend` folder. If you get an error, try running the `express-generator` package as an administrator.

Now that we have finished setting up the project, we can start working on the code. Next, we will start with creating the GraphQL backend.

Creating a GraphQL API with Express

To start the shopping cart system project, we first create a GraphQL API with Express. We start with the backend since we need it for both frontends. To get started, we have to add a few libraries that are needed to manipulate the SQLite database and add authentication to our app. Also, we need the library to enable **Cross-Origin Resource Sharing** (**CORS**) in our app.

CORS is a way to let us make requests from the browser to an endpoint hosted in a different domain from where the frontend is hosted.

To make our Express app accept GraphQL requests, we use the `graphql` and `express-graphql` libraries. To install both, we run the following command:

```
npm i cors jsonwebtoken sqlite3 express-graphql graphql
```

After installing the packages, we are ready to work on the code.

Working with resolver functions

First, we work on the resolver functions. To add them, we first add a `resolvers` folder into the `backend` folder. Then, we can work on the resolver for authentication. In the `resolvers` folder, we create an `auth.js` file and write the following code:

```
const jwt = require('jsonwebtoken');

module.exports = {
  login: ({ user: { username, password } }) => {
    if (username === 'admin' && password === 'password') {
      return { token: jwt.sign({ username }, 'secret') }
    }
    throw new Error('authentication failed');
  }
}
```

The `login` method is a resolver function. It takes the `user object` property with the `username` and `password` properties, and we use these to check for the credentials. We check if the username is `'admin'` and the password is `'password'`. If the credentials are correct, then we issue the token. Otherwise, we throw an error, which will be returned as an error response by the `/graphql` endpoint.

Adding resolvers for the order logic

We next add the resolvers for the order logic. In the `resolvers` folder, we add the `orders.js` file. Then, we work on the resolver function to get the order data. The order data has information about the order itself and also about what has been bought by the customer. To add the resolvers, we write the following code:

```javascript
const sqlite3 = require('sqlite3').verbose();

module.exports = {
  getOrders: () => {
    const db = new sqlite3.Database('./db.sqlite');
    return new Promise((resolve, reject) => {
      db.serialize(() => {
        db.all(`
          SELECT *,
            orders.name AS purchaser_name,
            shop_items.name AS shop_item_name
          FROM orders
          INNER JOIN order_shop_items ON orders.order_id =
            order_shop_items.order_id
          INNER JOIN shop_items ON
            order_shop_items.shop_item_id = shop_items.
              shop_item_id
        `, [], (err, rows = []) => {
          ...
        });
      })
      db.close();
    })
  },
  ...
}
```

We open the database with the sqlite3.Database constructor, with the path to the database. Then, we return a promise that queries all the orders with the items that the custom bought. The orders are in the orders table. The store inventory items are stored in the shop_items table, and we have the order_shop_items table to link the order and the items bought.

We make a `select` query with the `db.all` method to get all the data, and we join all the related tables with an `inner join` to get the related data in the other tables. In the callback, we write the following code to loop through the rows to create the `order` object:

```
const sqlite3 = require('sqlite3').verbose();

module.exports = {
  getOrders: () => {
    const db = new sqlite3.Database('./db.sqlite');
    return new Promise((resolve, reject) => {
      db.serialize(() => {
        db.all(`

          ...

        `, [], (err, rows = []) => {
          if (err) {
            reject(err)
...
          const orderArr = Object.values(orders)
          for (const order of orderArr) {
            order.ordered_items = rows
              .filter(({ order_id }) => order_id ===
                order.order_id)
              .map(({ shop_item_id, shop_item_name: name,
                price, description }) => ({
                shop_item_id, name, price, description
              }))
          }
          resolve(orderArr)
        });
      })
      db.close();
    })
  },
  ...
}
```

This lets us remove duplicate order entries in the rows. The key is the `order_id` value, and the value is the order data itself. Then, we get all order values with the `Object.values` method. We assign the returned array to the `orderArr` variable. Then, we loop through the `orderArr` array to get all the shop items that were ordered from the original row's array with the `filter` method, to look up the items by `order_id`. We call map to extract the shop item data of the order from the row.

We call `resolve` on the data to return it as a response from the `/graphql` endpoint. In the first few lines of the callback, we call `reject` when `err` is truthy so that we can return the error to the user, if there is one.

Finally, we call `db.close()` to close the database once we're done. We can do this at the end, since we used `db.serialize` to run all the statements in the `serialize` callback in a series so that the **Structured Query Language (SQL)** code could be run in sequence.

Adding an order

We add a resolver function to add an order. To do this, we write the following code:

```
const sqlite3 = require('sqlite3').verbose();


module.exports = {
  ...
  addOrder: ({ order: { name, address, phone, ordered_items:
    orderedItems } }) => {
    const db = new sqlite3.Database('./db.sqlite');
    return new Promise((resolve) => {
      db.serialize(() => {
        const orderStmt = db.prepare(`
        INSERT INTO orders (
          name,
          address,
          phone
          ...
              shop_item_id: shopItemId
            } = orderItem
            orderShopItemStmt.run(orderId, shopItemId)
          }
          orderShopItemStmt.finalize()
      })
```

```
            resolve({ status: 'success' })
            db.close();
        });
      })
    })
  },
  ...
}
```

We get the request payload for the order, with the variables we destructured within the argument. We open the database the same way, and we start with the same promise code and the `db.serialize` call, but inside it we create a prepared statement with the `db.prepare` method. We issue an INSERT statement to add the data to the order entry.

Then, we call `run` with the variable values we want to insert, to run the SQL statement. Prepared statements are good since all the variable values we passed into `db.run` are sanitized to prevent SQL injection attacks. Then, we call `finalize` to commit the transaction.

Next, we get the ID value of the row that has just been inserted into the `orders` table with the `db.all` call, with the SELECT statement. In the callback of the `db.all` method, we get the returned data and destructure `orderId` from the returned data.

Then, we create another prepared statement to insert the data for the shop items that were bought into the `order_shop_items` table. We just insert `order_id` and `shop_item_id` to link the order to the shop item bought.

We loop through the `orderedItems` array and call `run` to add the entries, and we call `finalize` to finalize all the database transactions.

Finally, we call `resolve` to return a success response to the client.

To finish off this file, we add the `removeOrder` resolver to let us remove orders from the database. To do this, we write the following code:

```
const sqlite3 = require('sqlite3').verbose();

module.exports = {
  ...
  removeOrder: ({ orderId }) => {
    const db = new sqlite3.Database('./db.sqlite');
    return new Promise((resolve) => {
```

```
        db.serialize(() => {
            const delOrderShopItemsStmt = db.prepare("DELETE FROM
                order_shop_items WHERE order_id = (?)");
            delOrderShopItemsStmt.run(orderId)
            delOrderShopItemsStmt.finalize();

            const delOrderStmt = db.prepare("DELETE FROM orders
                WHERE order_id = (?)");
            delOrderStmt.run(orderId)
            delOrderStmt.finalize();

            resolve({ status: 'success' })
        })
        db.close();
    })
},
}
```

We call `db.serialize` and `db.prepare` in the same way as we did before. The only difference is that we are issuing `DELETE` statements to delete everything with the given `order_id` in the `order_shop_items` and `orders` tables. We need to delete items from the `order_shop_items` table first since the order is still being referenced there.

Once we get rid of all the references of the order outside the `orders` table, we can delete the order itself in the `orders` table.

Getting the shop items

We create a `shopItems.js` file in the `resolvers` folder to hold the resolver functions for getting and setting the shop items. First, we start with a resolver function to get all the shop items. To do this, we write the following code:

```
const sqlite3 = require('sqlite3').verbose();

module.exports = {
  getShopItems: () => {
    const db = new sqlite3.Database('./db.sqlite');
    return new Promise((resolve, reject) => {
      db.serialize(() => {
```

```
            db.all("SELECT * FROM shop_items", [], (err, rows =
              []) => {
                if (err) {
                  reject(err)
                }
                resolve(rows)
              });
            })
          db.close();
        })
      },
      ...
    }
```

We call db.serialize and db.all, as we did before. We just get all the shop_items
entries with the query and we call resolve to return the selected data as a response to
the client.

Adding a resolver function to add a shop item

We will now add a resolver function to add a shop item. To do this, we write the
following code:

```
const sqlite3 = require('sqlite3').verbose();

module.exports = {
  ...
  addShopItem: ({ shopItem: { name, description, image_url:
    imageUrl, price } }) => {
    const db = new sqlite3.Database('./db.sqlite');
    return new Promise((resolve) => {
      db.serialize(() => {
        const stmt = db.prepare(`
          INSERT INTO shop_items (
            name,
            description,
            image_url,
            price
```

```
            ) VALUES (?, ?, ?, ?)
            `

        );
        stmt.run(name, description, imageUrl, price)
        stmt.finalize();
        resolve({ status: 'success' })
      })
      db.close();
    })
  },
  ...
}
```

We issue an INSERT statement to insert an entry, with the values destructured from the parameter.

Finally, we add the removeShopItem resolver by writing the following code to let us remove an entry from the shop_items table by its ID:

```
const sqlite3 = require('sqlite3').verbose();

module.exports = {
  ...
  removeShopItem: ({ shopItemId }) => {
    const db = new sqlite3.Database('./db.sqlite');
    return new Promise((resolve) => {
      db.serialize(() => {
        const stmt = db.prepare("DELETE FROM shop_items WHERE
          shop_item_id = (?)");
        stmt.run(shopItemId)
        stmt.finalize();
        resolve({ status: 'success' })
      })
      db.close();
    })
  },
}
```

Mapping resolvers to queries and mutations

We need to map the resolvers to queries and mutations so that we can call them when making GraphQL API requests. To do this, we go to the app.js file and add a few things. We will also add some middleware so that we can enable cross-domain communication and token checks for some requests. To do this, we start by writing the following code:

```
const createError = require('http-errors');
const express = require('express');
const path = require('path');
const cookieParser = require('cookie-parser');
const logger = require('morgan');
const { graphqlHTTP } = require('express-graphql');
const { buildSchema } = require('graphql');
const cors = require('cors')
const shopItemResolvers = require('./resolvers/shopItems')
const orderResolvers = require('./resolvers/orders')
const authResolvers = require('./resolvers/auth')
const jwt = require('jsonwebtoken');
```

We import everything we need with the require function. We can replace all the stuff at the top of the file with the preceding code block. We import the resolvers, the CORS middleware, the GraphQL library items, and the jsonwebtoken module.

Next, we create the schema for our GraphQL API by calling the buildSchema function. To do this, we write the following code:

```
...
const schema = buildSchema(`
  type Response {
    status: String
  }
  ...
  input Order {
    order_id: Int
    name: String
    address: String
    phone: String
    ordered_items: [ShopItem]
  }
```

```
...
type Query {
    getShopItems: [ShopItemOutput],
    getOrders: [OrderOutput]
}

type Mutation {
    addShopItem(shopItem: ShopItem): Response
    removeShopItem(shopItemId: Int): Response
    addOrder(order: Order): Response
    removeOrder(orderId: Int): Response
    login(user: User): Token
}
`);
...
```

The full schema definition can be found at `https://github.com/PacktPublishing/-Vue.js-3-By-Example/blob/master/Chapter07/backend/app.js`.

We have the `type` keyword to define a data type for a response, and we have the `Response` and `Token` types to use as the response. The `express-graphql` library will check the structure of the response against what is specified in the data type, so whatever query or mutation that returns data with the `Response` type should have a `status` `string` property. This is optional since we don't have an exclamation mark after the string.

The `input` keyword lets us define an `input` type. An `input` type is used for specifying the data structure of request payloads. They are defined the same way as `output` types with a list of properties, with their data type after the colon.

We can nest one data type in another, as we did with the `ordered_items` property in the `OrderOutput` type. We specify that it holds an array of objects with the `ShopItemOutput` data type. Likewise, we specify a similar data type for the `ordered_items` property in the `Order` data type. The square brackets indicate that the data type is an array.

`Query` and `Mutation` are special data types that let us add the resolver names before the colon and the data type of the output after the colon. The `Query` type specifies the queries, and the `Mutation` type specifies the mutations.

Next, we specify the `root` object with all the resolvers added to it, by writing the following code:

```
const root = {
    ...shopItemResolvers,
    ...orderResolvers,
    ...authResolvers
}
```

We just put all the resolvers we imported into the `root` object and we spread all the entries into the `root` object to merge them all into one object.

Then, we add `authMiddleware` to add an authentication check for some GraphQL requests. To do this, we write the following code:

```
const authMiddleware = (req, res, next) => {
  const { query } = req.body
  const token = req.get('authorization')
  const requiresAuth = query.includes('removeOrder') ||
    query.includes('removeShopItem') ||
    query.includes('addShopItem')
  if (requiresAuth) {
    try {
      jwt.verify(token, 'secret');
      next()
      return
    } catch (error) {
      res.status(401).json({})
      return
    }
  }
  next();
}
```

We get the `query` property from the JSON request payload to check which query or mutation the GraphQL request is invoking. Then, we get the `authorization` header with the `req.get` method. Next, we define a `requiresAuth` Boolean variable to check if the client is making requests that invoke the restricted queries or mutations.

If that is `true`, we call `jwt.verify` to verify the token with the secret. If it's valid, then we call `next` to proceed to the `/graphql` endpoint. Otherwise, we return a `401` response. If a `query` or `mutation` property doesn't need authentication, then we just call `next` to proceed to the `/graphql` endpoint.

Adding the middleware

Next, we add all the middleware we need to enable cross-domain communication, and add the `/graphql` endpoint to accept GraphQL requests. To do this, we write the following code:

```
...
const app = express();
app.use(cors())
app.use(logger('dev'));
app.use(express.json());
app.use(express.urlencoded({ extended: false }));
app.use(cookieParser());
app.use(authMiddleware)
app.use('/graphql', graphqlHTTP({
  schema,
  rootValue: root,
  graphiql: true,
}));
...
```

We write the following line of code to enable cross-domain communication:

```
app.use(cors())
```

The following line of code lets us accept JSON requests, which we also need for accepting GraphQL requests:

```
app.use(express.json());
```

The following line of code adds the authentication check to restricted GraphQL queries:

```
app.use(authMiddleware)
```

The preceding line of code must be added before the following code block:

```
app.use('/graphql', graphqlHTTP({
  schema,
  rootValue: root,
  graphiql: true,
}));
```

This way, the authentication check is done before the GraphQL request can be made. Finally, the following code block adds a /graphql endpoint to let us accept GraphQL requests:

```
app.use('/graphql', graphqlHTTP({
  schema,
  rootValue: root,
  graphiql: true,
}));
```

The grapgqlHTTP function returns a middleware after we pass in a bunch of options. We set the schema for the GraphQL API. The rootValue property has an object with all the resolvers. The resolver names should match the names specified in the Query and Mutation types. The graphiql property is set to true so that we can use the GraphiQL web app available when we go to the /graphql page in the browser.

To test authenticated endpoints, we can use the ModHeader extension available with Chrome and Firefox to add the authentication header with the token to the request headers. Then, we can test authenticated GraphQL requests easily.

> **Note**
>
> The extension can be downloaded from https://chrome.
> google.com/webstore/detail/modheader/
> idgpnmonknjnojddfkpgkljpfnnfcklj?hl=en for Chromium
> browsers and https://addons.mozilla.org/en-CA/firefox/
> addon/modheader-firefox/?utm_source=addons.
> mozilla.org&utm_medium=referral&utm_content=search
> for Firefox.

The following screenshot shows what the GraphiQL interface looks like. We also have the `ModHeader` extension to let us add the headers required to make authenticated requests at the top right of the screen:

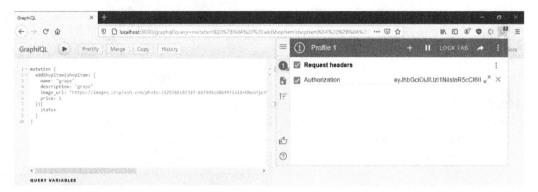

Figure 7.1 – GraphiQL with ModHeader extension

Next, we create a `db.sql` script to let us create the database we need to use, by writing the following code:

```sql
DROP TABLE IF EXISTS order_shop_items;
DROP TABLE IF EXISTS orders;
DROP TABLE IF EXISTS shop_items;

CREATE TABLE shop_items (
  shop_item_id INTEGER NOT NULL PRIMARY KEY,
  name TEXT NOT NULL,
  description TEXT NOT NULL,
  price NUMBER NOT NULL,
  image_url TEXT NOT NULL
);

CREATE TABLE orders (
  order_id INTEGER NOT NULL PRIMARY KEY,
  name TEXT NOT NULL,
  address TEXT NOT NULL,
  phone TEXT NOT NULL
);
```

```
CREATE TABLE order_shop_items (
  order_id INTEGER NOT NULL,
  shop_item_id INTEGER NOT NULL,
  FOREIGN KEY (order_id) REFERENCES orders(order_id)
  FOREIGN KEY (shop_item_id) REFERENCES
  shop_items(shop_item_id)
);
```

We create the tables that we used in the resolvers script. TEXT lets us store text in a column; INTEGER lets us store integers; FOREIGN KEY specifies a foreign key that references a column specified in the table and column after REFERENCES; NOT NULL makes a column required; DROP TABLE IF EXISTS drops a table, if it exists; CREATE TABLE creates a table; PRIMARY KEY specifies the primary key column.

Creating a SQLite database

To create and manipulate a SQLite database, we use the **DB Browser for SQLite (DB4S)** program, which we can download from https://sqlitebrowser.org/. This program works with Windows, Mac, and Linux. Then, we can click on **New Database** and save the db.sqlite database in the backend folder so that the backend can access the database. Then, in the **Execute SQL** tab, we paste in the script to add the tables to the database. For changes for the database to be written to disk, you have to save them. To do this, click on the **File** menu and then click on **Write Changes**. We can also press the *Ctrl + S* keyboard combination to save the changes.

Finally, in package.json, we change the start script by writing the following code:

```
{
  ...
  "scripts": {
    "start": "nodemon ./bin/www"
  },
  ...
}
```

We switch nodemon so that the app will restart when we change the code and save it. We run npm I -g nodemon to install nodemon globally.

Now that we are done with the backend, we can move on to the frontend so that we have a complete shopping cart system.

Creating the admin frontend

Now that we have the backend app done, we can move on to working on the frontend. Since we have already created the Vue 3 project earlier for the admin frontend in the admin-frontend folder, we just have to install packages that we need and then work on the code. We need the graphql-request GraphQL package and the GraphQL client library, and the VeeValidate, Vue Router, Axios, and Yup packages.

To install them, we run the following command in the admin-frontend folder:

```
npm i vee-validate@next vue-router@4 yup graphql graphql-
request
```

After installing the packages, we can start working on the code.

Working with components

First, we start working on the components. In the components folder, we add the TopBar component into the components/TopBar.vue file to hold the route links and the **Log Out** button by writing the following code:

```
<template>
  <p>
    <router-link to="/orders">Orders</router-link>
    <router-link to="/shop-items">Shop Items</router-link>
    <button @click="logOut">Log Out</button>
  </p>
</template>

<script>
export default {
  name: "TopBar",
  methods: {
    logOut() {
      localStorage.clear();
```

```
        this.$router.push("/");
      },
    },
  };
</script>

<style scoped>
a {
  margin-right: 5px;
}
</style>
```

We add the Vue Router `router-link` components to let the admin user click on them to go to different pages.

The **Log Out** button runs the `logOut` method when it is clicked to clear local storage with `localStorage.clear` and redirects back to the login page with `this.$router.push`. The / path will map to the login page, as we will see later.

Next, in the `src/plugins` folder, we add the `router.js` file. To do this, we write the following code:

```
import { createRouter, createWebHashHistory } from 'vue-router'
import Login from '@/views/Login'
import Orders from '@/views/Orders'
import ShopItems from '@/views/ShopItems'

const beforeEnter = (to, from, next) => {
  try {
    const token = localStorage.getItem('token')
    if (to.fullPath !== '/' && !token) {
      return next({ fullPath: '/' })
    }
    return next()
  } catch (error) {
    return next({ fullPath: '/' })
  }
}
```

```
const routes = [
  { path: '/', component: Login },
  { path: '/orders', component: Orders, beforeEnter },
  { path: '/shop-items', component: ShopItems, beforeEnter },
]

const router = createRouter({
  history: createWebHashHistory(),
  routes,
})

export default router
```

We added the `beforeEnter` route guard to check if the authentication token is stored in local storage. If it is stored already and we are going to an authenticated route, then we proceed to the page by calling `next` with no arguments. Otherwise, we redirect back to the login page by calling `next` with an object with the `fullPath` property set to `'/'`. If there is any error, we also go back to the login page.

Next, we have the `routes` array with the route mappings. We map the path to the component so that when we type in the **Uniform Resource Locator** (**URL**) in the browser or click a router link on the page, we go to the page we mapped to. We add the `beforeEnter` route guard to the route that requires authentication.

Then, we call `createRouter` to create the `router` object, and we call `createWebHashHistory` to use hash mode. With hash mode, the hostname and the rest of the URL will be separated by a # sign. We also add the `routes` array into the object we pass into `createRouter`, to add the route mappings.

Then, we export the `router` object so that we can add it to our app later.

Next, we create the login page component. To do this, we create the `views` folder, add the `Login.vue` file to it, and then write the following code:

```
<template>
  <h1>Admin Login</h1>
  <Form :validationSchema="schema" @submit="submitForm">
    <div>
      <label for="name">Username</label>
      <br />
      <Field name="username" type="text"
```

```
                placeholder="Username" />
            <ErrorMessage name="username" />
        </div>
        <br />
        <div>
            <label for="password">Password</label>
            <br />
            <Field name="password" placeholder="Password"
              type="password" />
            <ErrorMessage name="password" />
        </div>
        <input type="submit" />
    </Form>
</template>
```

We add the `Form` component with the `validationSchema` prop set to the `yup` schema. We listen for the `submit` event, which is emitted when all the fields are valid, and we click the **Submit** button. The `submitForm` method will have the form field values we entered, and the `Field` component lets us create a form field.

`ErrorMessage` displays the error message with the form field. If the `name` prop values of `Field` and `ErrorMessage` match, then any form validation for the field with the given name will be automatically displayed. The `placeholder` attribute lets us add a form placeholder, and the `type` attribute sets the `form` input type.

Next, we add the script portion of our component. To do this, we write the following code:

```
<script>
import { GraphQLClient, gql } from "graphql-request";
import * as yup from "yup";
import { Form, Field, ErrorMessage } from "vee-validate";


const APIURL = "http://localhost:3000/graphql";
const graphQLClient = new GraphQLClient(APIURL, {
  headers: {
    authorization: "",
  },
});
const schema = yup.object({
```

```
    name: yup.string().required(),
    password: yup.string().required(),
});

. . .
</script>
```

We create the GraphQL client object with the `GraphQLClient` constructor. This takes the GraphQL endpoint URL and various options that we can pass in. We will use it to pass in the required request headers in components that require authentication.

The `schema` variable holds the yup validation schema, which has the `name` and `password` fields. Both fields are strings and they are both required, as indicated by the method calls. The property names must match the `name` prop value of the `Field` and `ErrorMessage` components for the validation to trigger for the field.

Adding the login logic and making our first GraphQL request

Next, we add the login logic by writing the following code:

```
<script>
. . .

export default {
    name: "Login",
    components: {
        Form,
        Field,
        ErrorMessage,
    },
    data() {
        return {
            schema,
        };
    },
. . .
            } = await graphQLClient.request(mutation, variables);
```

```
            localStorage.setItem("token", token);
            this.$router.push('/orders')
        } catch (error) {
            alert("Login failed");
        }
    },
  },
};
</script>
```

We register the `Form`, `Field`, and `ErrorMessage` components imported from the VeeValidate package. We have the `data` method, which returns an object with the schema so that we can use it in the template. Finally, we have the `submitForm` method, to get the `username` and `password` values from the `Field` components and make the login mutation GraphQL request.

We pass the `$username` and `$password` values into the parentheses to pass them into our mutation. The values will be obtained from the `variablesvariables` object, which we pass into the `graphQLClient.request` method. If the request is successful, we then get back the token from the request. Once we get the token, we put it in `localStorage.setItem` to put it into local storage.

The `gql` tag is a function that lets us convert the string into a query JSON object that can be sent to the server.

If the login request failed, we then display an alert. The following screenshot shows the login screen:

Admin Login

Username

Username

Password

Password

Submit

Figure 7.2 – Admin login screen

Creating the orders page

Next, we create an orders page by creating a `views/Orders.vue` file. To do this, we update the following code:

```
<template>
 <TopBar />
 <h1>Orders</h1>
 <div v-for="order of orders" :key="order.order_id">
   <h2>Order ID: {{ order.order_id }}</h2>
   <p>Name: {{ order.name }}</p>
   <p>Address: {{ order.address }}</p>
   <p>Phone: {{ order.phone }}</p>
   <div>
     <h3>Ordered Items</h3>
     <div
       v-for="orderedItems of order.ordered_items"
       :key="orderedItems.shop_item_id"
     >
       <h4>Name: {{ orderedItems.name }}</h4>
       <p>Description: {{ orderedItems.description }}</p>
       <p>Price: ${{ orderedItems.price }}</p>
```

```
        </div>
      </div>
      <p>
        <b>Total: ${{ calcTotal(order.ordered_items) }}</b>
      </p>
      <button type="button" @click="deleteOrder(order)">Delete
        Order</button>
    </div>
  </template>
```

We add `TopBar` and loop through the orders with `v-for` to render the entries. We also loop through `ordered_items`. We show the total price of the ordered items with the `calcTotal` method. We also have the **Delete Order** button, which calls the `deleteOrder` method when we click it. The `key` prop must be specified so that Vue 3 can identify the items.

Next, we create a script with the GraphQL client by writing the following code:

```
<script>
import { GraphQLClient, gql } from "graphql-request";
import TopBar from '@/components/TopBar'
const APIURL = "http://localhost:3000/graphql";
const graphQLClient = new GraphQLClient(APIURL, {
  headers: {
    authorization: localStorage.getItem("token"),
  },
});


...
</script>
```

This is different from the login page since we set the authorization header to the token we obtained from local storage. Next, we create the component object by writing the following code:

```
<script>
...

export default {
  name: "Orders",
  components: {
    TopBar
...
          {
              getOrders {
                order_id
                name
                address
                phone
                ordered_items {
                  shop_item_id
                  name
                  description
                  image_url
                  price
                }
              }
          }
        `;
...
        await graphQLClient.request(mutation, variables);
        await this.getOrders();
      },
    },
};
</script>
```

We register the `TopBar` component with the `components` property. We have the `data` method to return an object with the `orders` reactive property. In the `beforeMount` hook, we call the `getOrders` method to get the orders when the component is mounting. The `calcTotal` method calculates the total price of all the ordered items by getting the price from all the `orderedItems` objects with `map` and then calling `reduce` to add all the prices together.

The `getOrders` method makes a GraphQL query request to get all the orders. We specify the fields we want to get with the request. We specify the fields for the nested objects we also want to get, so we do the same with `ordered_items`. Only the fields that are specified will be returned.

Then, we call `graphQlClient.request` with the query to make the query request, and assign the data returned to the `orders` reactive property.

The `deleteOrder` method takes an `order` object and makes a `removeOrder` mutation request to the server. `orderId` is in the variables, so the correct order will be deleted. We call `getOrders` to get the latest orders again after deleting them.

The following screenshot shows the orders page that the admin sees:

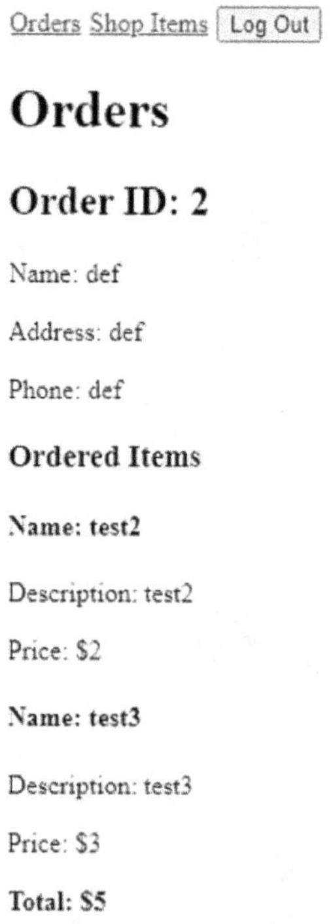

Orders Shop Items [Log Out]

Orders

Order ID: 2

Name: def

Address: def

Phone: def

Ordered Items

Name: test2

Description: test2

Price: $2

Name: test3

Description: test3

Price: $3

Total: $5

Figure 7.3 – Orders page: admin view

Now that we have added the orders page, we will move on to add a page to let admins add and remove items they want to sell in the shop.

Adding and removing shop items for sale

Next, we add a shop items page to let us add and remove shop items. To do this, we start with the template. We render the shop items by writing the following code:

```
<template>
  <TopBar />
  <h1>Shop Items</h1>
  <button @click="showDialog = true">Add Item to Shop</button>
  <div v-for="shopItem of shopItems"
    :key="shopItem.shop_item_id">
    <h2>{{ shopItem.name }}</h2>
    <p>Description: {{ shopItem.description }}</p>
    <p>Price: ${{ shopItem.price }}</p>
    <img :src="shopItem.image_url" :alt="shopItem.name" />
    <br />
    <button type="button" @click="deleteItem(shopItem)">
      Delete Item from Shop
    </button>
  </div>

  ...
</template>
```

We add the `TopBar` component as we did before and we render `shopItems`, as we did with the orders.

Next, we add a dialog box with the HTML dialog element to let us add shop items. To do this, we write the following code:

```
<template>
  ...

  <dialog :open="showDialog" class="center">
    <h2>Add Item to Shop</h2>
    <Form :validationSchema="schema" @submit="submitForm">
```

```
    <div>
...
      <Field name="imageUrl" type="text" placeholder=" Image
        URL" />
      <ErrorMessage name="imageUrl" />
    </div>
    <br />
    <div>
      <label for="price">Price</label>
      <br />
      <Field name="price" type="text" placeholder="Price" />
      <ErrorMessage name="price" />
    </div>
    <br />
    <input type="submit" />
    <button @click="showDialog = false" type="button">
      Cancel</button>
  </Form>
  </dialog>
</template>
```

We set the open prop to control when the dialog box is opened, and we set the class to center so that we can apply styles to center the dialog box and display it above the rest of the page later.

Inside the dialog box, we have the form created in the same way as with the login page. The only difference is the fields that are in the form. At the bottom of the form, we have a **Cancel** button to set the showDialog reactive property to false to close the dialog, since it's set as the value of the open prop.

Next, we create the script with the GraphQL client and form validation schema (as we did before), as follows:

```
<script>
import { GraphQLClient, gql } from "graphql-request";
import * as yup from "yup";
import TopBar from "@/components/TopBar";
import { Form, Field, ErrorMessage } from "vee-validate";
```

```
const APIURL = "http://localhost:3000/graphql";
const graphQLClient = new GraphQLClient(APIURL, {
  headers: {
    authorization: localStorage.getItem("token"),
  },
});
const schema = yup.object({
  name: yup.string().required(),
  description: yup.string().required(),
  imageUrl: yup.string().required(),
  price: yup.number().required().min(0),
});


...
</script>
```

Then, we add the `component options` object by writing the following code:

```
<script>
...
export default {
  name: "ShopItems",
  components: {
    Form,
    Field,
    ErrorMessage,
    TopBar,
  },
  data() {
    return {
      shopItems: [],
      showDialog: false,
      schema,
    };
  },
  beforeMount() {
    this.getShopItems();
```

```
    },
    ...
};
</script>
```

We register components and create a `data` method to return the reactive properties we use. The `beforeMount` hook calls the `getShopItems` method to get the shop items from the API.

Next, we add the `getShopItems` method by writing the following code:

```
<script>
...
export default {
  ...
  methods: {
    async getShopItems() {
      const query = gql`
        {
          getShopItems {
            shop_item_id
            name
            description
            image_url
            price
          }
        }
      `;
      const { getShopItems: data } = await
        graphQLClient.request(query);
      this.shopItems = data;
    },
    ...
  },
};
</script>
```

We just make a `getShopItems` query request to get the data with the fields in the braces returned.

Next, we add the `submitForm` method to make a mutation request to add a shop item entry, by writing the following code:

```
<script>
...
export default {
  ...
  methods: {
    ...
    async submitForm({ name, description, imageUrl, price:
      oldPrice }) {
      const mutation = gql`
        mutation addShopItem(
          $name: String
          $description: String
          $image_url: String
          $price: Float
        ) {
...
          description,
          image_url: imageUrl,
          price: +oldPrice,
      };
      await graphQLClient.request(mutation, variables);
      this.showDialog = false;
      await this.getShopItems();
    },
    ...
  },

};
</script>
```

We get all the form field values by destructuring the object in the parameter, then we call `graphQLClient.request` to make the request with the variables set from the destructured properties from the parameter. We convert `price` to a number since `price` should be a float, according to the schema we created in the backend.

Once the request is done, we set `showDialog` to `false` to close the dialog and we call `getShopItems` again to get the shop items.

The last method we'll add is the `deleteItem` method. The code for this can be seen in the following snippet:

```
<script>
...
export default {
  ...
  methods: {
    ...
    async deleteItem({ shop_item_id: shopItemId }) {
      const mutation = gql`
        mutation removeShopItem($shopItemId: Int) {
          removeShopItem(shopItemId: $shopItemId) {
            status
          }
        }
      `;
      const variables = {
        shopItemId,
      };
      await graphQLClient.request(mutation, variables);
      await this.getShopItems();
    },
    ...
  },
};
</script>
```

We make the `removeShopItem` mutation request to delete a shop item entry. When the request is done, we call `getShopItems` again to get the latest data.

The admin's view of the shop items page can be seen in the following screenshot:

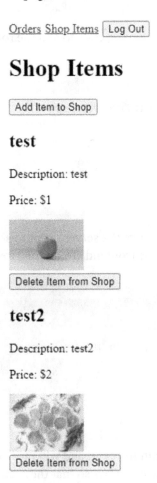

Figure 7.4 – Shop items page: admin view

In `src/App.vue`, we write the following code to add the `router-view` component to show the route component content:

```
<template>
  <router-view></router-view>
</template>

<script>
```

```
export default {
  name: "App",
};
</script>
```

In `src/main.js`, we write the following code to add the router to our app:

```
import { createApp } from 'vue'
import App from './App.vue'
import router from '@/plugins/router'

const app = createApp(App)
app.use(router)
app.mount('#app')
```

Finally, in `package.json`, we change the server script to serve the app from a different port so that it won't conflict with the frontend. To do this, we write the following code:

```
{
  ...
  "scripts": {
    "serve": "vue-cli-service serve --port 8090",
    ...
  },
  ...
}
```

We are now finished with the admin frontend and will move on to the final part of this project, which is a frontend for the customer so that they can order items.

Creating the customer frontend

Now that we have finished with the admin frontend, we finish this chapter's project by creating the customer's frontend. This is similar to the admin frontend except that there is no authentication required to use it.

We start by installing the same packages that we installed for the admin frontend. So, we go to the `frontend` folder and run the following command to install all the packages:

```
npm i vee-validate@next vue-router@4 yup vuex@4 vuex-
persistedstate@ ^4.0.0-beta.3 graphql graphql-request
```

We need Vuex with the `Vuex-Persistedstate` plugin to store the shopping cart items. The rest of the packages are the same as for the admin frontend.

Creating the plugins folder

We create a `plugins` folder in the `src` folder and add the routes by creating the `router.js` file in the folder and writing the following code:

```js
import { createRouter, createWebHashHistory } from 'vue-router'
import Shop from '@/views/Shop'
import OrderForm from '@/views/OrderForm'
import Success from '@/views/Success'

const routes = [
    { path: '/', component: Shop },
    { path: '/order-form', component: OrderForm },
    { path: '/success', component: Success },
]

const router = createRouter({
  history: createWebHashHistory(),
  routes,
})
```

Next, we create our Vuex store by creating the `src/plugins/vuex.js` file and then writing the following code:

```js
import { createStore } from "vuex";
import createPersistedState from "vuex-persistedstate";

const store = createStore({
  state() {
    return {
      cartItems: []
```

```
        }
    },
    getters: {
        cartItemsAdded(state) {
            return state.cartItems
        }
    },
    mutations: {
        addCartItem(state, cartItem) {
            const cartItemIds = state.cartItems.map(c =>
                c.cartItemId).filter(id => typeof id === 'number')
            state.cartItems.push({
...
            state.cartItems = []
        }
    },
    plugins: [createPersistedState({
        key: 'cart'
    })],
});
```

```
export default store
```

We call `createStore` to create the Vuex store. In the object that we pass into `createStore`, we have the `state` method to return the `cartItems` state initialized to an array. The `getters` property has an object with the `cartItemsAdded` method to return the `cartItems` state value.

In the `mutations` property object, we have the `addCartItem` method to call `state.cartItems.push` to add a `cartItem` value to the `cartItems` state. We get the existing cart item IDs with the `map` and `filter` method. We only want the numeric ones. The ID for the new cart item would be the highest one from the `cartItemIds` array plus 1.

The `removeCartItem` method lets us call `splice` to remove a cart item by index, and `clearCart` resets the `cartItems` state to an empty array.

Finally, we set the `plugins` property to an object with the `createPersistedState` function, to create a Vuex-Persistedstate plugin to store the `cartItems` state to local storage. The `key` value is the key to store the `cartItem` values under. Then, we export the store so that we can add it to our app later.

Creating the order form page

Next, we create an order form page. This has a form to let customers enter their personal information and edit the cart. To create it, we create a `src/views` folder if one doesn't already exist, and then we create an `OrderForm.vue` component file. We start by writing the following template code:

```
<template>
  <h1>Order Form</h1>
  <div v-for="(cartItem, index) of cartItemsAdded"
    :key="cartItem.cartItemId">
    <h2>{{ cartItem.name }}</h2>
    <p>Description: {{ cartItem.description }}</p>
    <p>Price: ${{ cartItem.price }}</p>
    <br />

...

      <Field name="phone" type="text" placeholder="Phone" />
      <ErrorMessage name="phone" />
    </div>
    <br />
    <div>
      <label for="address">Address</label>
      <br />
      <Field name="address" type="text" placeholder="Address"
        />
      <ErrorMessage name="address" />
    </div>
    <br />
    <input type="submit" />
  </Form>
</template>
```

We have similar forms to those for the admin frontend. We use the same `Form`, `Field`, and `ErrorMessage` components from VeeValidate.

We loop through the cart items with `v-for` to render them onto the screen. They're retrieved from local storage via `Vuex-Persistedstate` with the `cartItemsAdded` getter.

Next, we create the script the same way by writing the following code:

```
<script>
import { GraphQLClient, gql } from "graphql-request";
import { mapMutations, mapGetters } from "vuex";
import { Form, Field, ErrorMessage } from "vee-validate";
import * as yup from "yup";


...
export default {
  name: "OrderForm",
  data() {
    return {
      schema,
    };
  },
  components: {
    Form,
    Field,
    ErrorMessage,
  },
  computed: {
    ...mapGetters(["cartItemsAdded"]),
  },
  ...
};
</script>
```

We create the GraphQL client and the validation schema, and we register the components in the same way as we did in the shop item page of the admin frontend. The only new thing is to call the `mapGetters` method to add the Vuex getters as a computed property of our component. We just pass in an array of strings with the name of the getters to map the computed properties to. Next, we add the methods by writing the following code:

```
<script>
...
export default {
  ...
  methods: {
    async submitOrder({ name, phone, address }) {
      const mutation = gql`
        mutation addOrder(
          $name: String
          $phone: String
          $address: String
          $ordered_items: [ShopItem]
...
            shop_item_id,
            name,
            description,
            image_url,
            price,,
          })
        ),
      };
      await graphQLClient.request(mutation, variables);
      this.clearCart();
      this.$router.push("/success");
    },
    ...mapMutations(["addCartItem", "removeCartItem",
      "clearCart"]),
  },
};
</script>
```

We have the `submitOrder` method that gets the inputted data from the order form and makes an `addOrder` mutation request to the server. In the `variables` object, we need to remove `cartItemId` from each `ordered_items` object so that it matches the `ShopItem` schema that we created in the backend. We can't have extra properties that aren't included in the schema in an object we send to the server.

Once the request succeeds, we call `clearCart` to clear the cart, and then we call `thus.$router.push` to go to the success page. The `mapMutation` method maps the mutations to methods in our component. The `clearCart` method is the same as the `clearCart` Vuex store mutation.

The following screenshot shows the admin view of the order form:

Order Form

test

Description: test

Price: $1

Remove From Cart

test2

Description: test2

Price: $2

Remove From Cart
Name
james

Phone
Phone

Address
Address

Submit

Figure 7.5 – Order form: admin view

Next, we create a `src/views/Shop.vue` file by writing the following code:

```html
<template>
  <h1>Shop</h1>
  <div>
    <router-link to="/order-form">Check Out</router-link>
  </div>
  <button type="button" @click="clearCart()">Clear Shopping
    Cart</button>
  <p>{{ cartItemsAdded.length }} item(s) added to cart.</p>
  <div v-for="shopItem of shopItems" :key="shopItem.
    shop_item_id">
    <h2>{{ shopItem.name }}</h2>
    <p>Description: {{ shopItem.description }}</p>
    <p>Price: ${{ shopItem.price }}</p>
    <img :src="shopItem.image_url" :alt="shopItem.name" />
    <br />
    <button type="button" @click="addCartItem(shopItem)">Add
        to Cart</button>
  </div>
</template>
```

We render the shop items with `v-for`, as we did with the other components. We also have a `router-link` component to render a link on the page.

We show the number of cart items added with the `cartItemsAdded` getter. The `clearCart` Vuex mutation method is called when we click on **Clear Shopping Cart**. Next, we add the script for the component by writing the following code:

```js
<script>
import { GraphQLClient, gql } from "graphql-request";
import { mapMutations, mapGetters } from "vuex";

const APIURL = "http://localhost:3000/graphql";
const graphQLClient = new GraphQLClient(APIURL);

...
    async getShopItems() {
```

```
        const query = gql`
          {
            getShopItems {
              shop_item_id
              name
              description
              image_url
              price
            }
          }
        `;
        const { getShopItems: data } = await
          graphQLClient.request(query);
        this.shopItems = data;
      },
      ...mapMutations(["addCartItem", "clearCart"]),
    },
  };
</script>
```

We create the GraphQL client the same way. In the component, we call `getShopItems` in the `beforeMount` hook to get the shopping cart items. We also call `mapMutations` to map the Vuex mutations we need into methods in our component.

Finally, we shrink the `img` elements to `100px` width by writing the following code:

```
<style scoped>
img {
  width: 100px;
}
</style>
```

Next, we create an order success page by creating a `src/views/Success.vue` file and writing the following code:

```
<template>
  <div>
    <h1>Order Successful</h1>
    <router-link to="/">Go Back to Shop</router-link>
```

```
      </div>
    </template>

    <script>
    export default {
      name: "Success",
    };
    </script>
```

The order success page just has some text and a link to go back to the shop's home page.

Next, in src/App.vue, we write the following code to add the router-view component to show the route pages:

```
    <template>
      <router-view></router-view>
    </template>

    <script>
    export default {
      name: "App",
    };
    </script>
```

In src/main.js, we add the following code to add the router and Vuex store to our app:

```
    import { createApp } from 'vue'
    import App from './App.vue'
    import router from '@/plugins/router'
    import store from '@/plugins/vuex'

    const app = createApp(App)
    app.use(router)
    app.use(store)
    app.mount('#app')
```

And finally, we change the port that the app project is served from by writing the following code:

```
{
    ...
    "scripts": {
        "serve": "vue-cli-service serve --port 8091",
        ...
    },
    ...
}
```

Our project is now complete.

We can run the frontend projects with `npm run serve` and the backend projects with `npm start`.

By working on the shopping cart project, we learned how to create GraphQL APIs, which are JSON APIs that can process GraphQL instructions via queries and mutations.

Summary

We can easily create a GraphQL API with Express and the `express-graphql` library. To make GraphQL HTTP requests easily, we use the `graphql-request` JavaScript GraphQL client, which works in the browser. This lets us set request options such as headers, the query to make, and variables that go with the query easily.

The `graphql-request` GraphQL client is used instead of a regular HTTP client to make requests to the backend from our Vue app. The `graphql-request` library lets us make GraphQL HTTP requests more easily than when using a regular HTTP client. With it, we can easily pass in GraphQL queries and mutations with variables.

A GraphQL API is created with a schema that maps to resolver functions. Schemas let us define all the data types of our input and output data so that we don't have to guess which data we have to send. If we send any invalid data, then we will get an error telling us exactly what is wrong with the request. We also have to specify the data fields that we want to return with our GraphQL queries, and only the fields that we specified are returned. This lets us return the data that we need to use, making it much more efficient.

Also, we can add authentication to a GraphQL API request with the usual token check before making requests to the /graphql endpoint.

We can easily test GraphQL requests with the GraphiQL interactive sandbox that lets us make the requests we want. To test authenticated requests, we can use the ModHeader extension to set the header so that we can make authenticated requests successfully.

In the next chapter, we will look at how to create a real-time chat app with Laravel and Vue 3.

8

Building a Chat App with Vue 3, Laravel, and Socket.IO

In the previous chapters, we have created frontend projects or full stack projects that only communicate via HTTP. There is no real-time communication between the frontend and the backend. Real-time communication is sometimes necessary if we need to communicate data from the server side to the client side and vice versa instantly. Without some real-time communication mechanism, there is no way to communicate from the server side to the client side without the client initiating the request. This is something that we can add easily with the Laravel framework and Socket.io.

In this chapter, we will take a look at the following topics:

- Creating the API endpoints with Laravel

- Setting up JWT authentication

- Creating the frontend to let users chat

Laravel is a backend web framework that is written with PHP. It is a comprehensive backend framework that includes processing HTTP requests, database manipulation, and real-time communication.

In this chapter, we will look at how to get all these parts working together so that we can create a chat app with Vue 3, Laravel, Laravel Echo Server, and Redis working together.

Technical requirements

To fully understand this chapter, the following is required:

- A basic understanding of PHP
- The ability to create basic apps with Vue components
- The ability to send and receive HTTP requests with the Axios HTTP client

The code for this chapter's project is available at `https://github.com/PacktPublishing/-Vue.js-3-By-Example/tree/master/Chapter08`.

Creating the API endpoints with Laravel

The first step to create our chat app is to create a backend app with Laravel. Creating the API with Laravel is the main thing that we have to learn for this chapter. This is something that we have not done before. It also means that we have to write code in PHP since Laravel is a PHP-based web framework. Therefore, you should learn some basic PHP syntaxes before reading this code. Like JavaScript and other object-oriented languages, they share similar concepts such as using objects, arrays, dictionaries, loops, classes, and other basic object-oriented programming concepts. Therefore, it should not be too different from JavaScript in terms of difficulty of learning.

Installing the required libraries

To create our API with Laravel, we don't have to create all the files ourselves, we just have to run a few commands and that will create all the files and configuration settings for us automatically. Before we create our API, we have to have PHP running. In Windows, the easiest way to add PHP to our Windows installation is to use XAMPP. We can download and install it by going to `https://www.apachefriends.org/download.html`. It is also available for macOS and Linux.

Once we install it, then we can create our Laravel API with **Composer**. Composer is a package manager for PHP that we will use to install more libraries later on. The easiest way to create a project is to create our project folder and then run the command to create the Laravel project after we go to the folder:

1. First, we create a project folder called `vue-example-ch8-chat-app` that will hold both the frontend and backend in their own separate folders.

2. Then, within this folder, we create the backend folder to house our Laravel project code files.

3. Now we go to the command line, then we go into `vue-example-ch8-chat-app` and then run `composer global require laravel/installer`.

This will install the Laravel installer, which will let us create our Laravel project. The locations at which the global libraries are located are as follows:

- **macOS**: `$HOME/.composer/vendor/bin`

- **Windows**: `%USERPROFILE%\AppData\Roaming\Composer\vendor\bin`

- **GNU / Linux Distributions**: `$HOME/.config/composer/vendor/bin` or `$HOME/.composer/vendor/bin`

We can also run `composer global about` to find out where the library files are located.

Once that is done, we create the scaffold with all the files and also include all the configuration files and install all the required libraries for us with one command.

We go into the `vue-example-ch8-chat-app` folder through the command line, and then we run `laravel new backend` to create the Laravel app in the backend folder. The Laravel installer will run and create the scaffolding for our Laravel. Also, Composer will install all the PHP libraries that we need to run Laravel. Once that is all done, we should have a full Laravel installation with all the files and configuration that we need to run our app.

Creating databases and migration files

Now, with the creation of the Laravel app and all the associated libraries being installed, we can work on the Laravel app to create our API. First, we create our database by creating some migration files. We need them to create the `chats` and `messages` tables. The `chats` table has the chat room data. And the `messages` table has the chat messages that are associated with a chat room. It will also have a reference to the user who sent the message.

We don't have to create a `users` table since that is created automatically when we create the Laravel app. Almost every app needs to hold user data so this is included automatically. With the Laravel scaffold, we can create users with a username, email, and password, and log in with the username and password for the user that we just created. Laravel also has the ability to send emails for user verification without having to add any code to do that.

To create the migrations, we run the following commands:

```
php artisan make:migration create_chats_table
php artisan make:migration create_messages_table
```

The preceding commands will create the migration files for us with the date and time prepended to the filename of the migration files. All the migration files are in the `database/migrations` folder. So we can go into this folder and open the files. In the one with the `create_chats_table` as the filename, we add the following code:

```php
<?php

use Illuminate\Database\Migrations\Migration;
use Illuminate\Database\Schema\Blueprint;
use Illuminate\Support\Facades\Schema;

class CreateChatsTable extends Migration
{
    public function up()
    {
        Schema::create('chats', function (Blueprint $table)
        {
            $table->id();
            $table->string('name');
            $table->timestamp('created_at')->useCurrent();
            $table->timestamp('updated_at')->useCurrent();
        });
    }

    public function down()
    {
```

```
            Schema::dropIfExists('chats');
    }
}
```

The preceding code will create the `chats` table. The `up()` method has the code that we want to run when we run our migration. The `down()` method has the method we run when we want to reverse the migration.

In the `up()` method, we call `Schema::create` to create the table. The `::` symbols indicate that the method is a static method. The first argument is the table name, and the second argument is a callback function that we add code to create the table with. The `$table` object has the `id()` method to create an `id` column. The `string()` method creates a `string` column with the column name in the argument. The `timestamp()` method lets us create a `timestamp` column with the given column name. The `useCurrent()` method lets us set the default value of the timestamp to the current date and time.

In the `down()` method, we have the `Schema::dropIfExists()` method to drop the table with the given name in the argument to drop the table if it exists.

A migration file must have a class that inherits from the `Migration` class for it to be used as a migration.

Likewise, in the migration file with the `create_message_table` name in the file name, we write the following:

```php
<?php

use Illuminate\Database\Migrations\Migration;
use Illuminate\Database\Schema\Blueprint;
use Illuminate\Support\Facades\Schema;

class CreateMessagesTable extends Migration
{
    public function up()
    {
        Schema::create('messages', function (Blueprint $table) {
            $table->id();
            $table->unsignedBigInteger('user_id');
```

```
        $table->unsignedBigInteger('chat_id');
        $table->string('message');
        $table->timestamp('created_at')->useCurrent();
        $table->timestamp('updated_at')->useCurrent();
        $table->foreign('user_id')->references('id')-
            >on('users');
        $table->foreign('chat_id')->references('id')-
            >on('chats');
    });
}


public function down()
{
    Schema::dropIfExists('messages');
}
}
```

The preceding file has the code to create the `messages` table. This table has more columns. We have the same `id` and `timestamp` columns as in the `chats` table, but we also have the `user_id` unsigned `integer` column to reference the ID of the user who posted the message and the `chat_id` unsigned `integer` column to reference an entry in the `chats` table to associate the message with the chat session that it is created in.

The `foreign()` method lets us specify what the `user_id` and `chat_id` columns are referencing in the users and `chats` tables respectively.

Configuring our database

Before we can run our migration, we have to configure our database that we will use to store the data for our backend. To do that, we create the `.env` file in the project's `root` folder by copying the `.env.example` file and then renaming it `.env`.

The `.env` file has many settings that we will need to run our Laravel app. To configure which database we will use, we run the following command to let us connect to a SQLite database:

```
DB_CONNECTION=sqlite
DB_DATABASE=C:\vue-example-ch8-chat-app\backend\db.sqlite
```

The full configuration file is at `https://github.com/PacktPublishing/-Vue.js-3-By-Example/blob/master/Chapter08/backend/.env.example`. We just copy its contents into the `.env` file in the same folder to use the configuration.

We use SQLite for simplicity in this chapter so that we can focus on creating the chat app with Vue 3. However, we should use a production-quality database that has better security and management capabilities if we are building a production app. The `DB_CONNECTION` environment variable has the database type we want to use, which is SQLite. In the `DB_DATABASE` setting, we specify the absolute path of our database file. Laravel will not create this file automatically for us, so we have to create it ourselves. To create the SQLite file, we can use the DB Browser for SQLite program. It supports Windows, macOS, and Linux so we can run this on all the popular platforms. You can download the program from `https://sqlitebrowser.org/dl/`. Once this is installed, just click on **New Database** at the top left and click on the **File** menu and click **Save** to save the database file.

Configuring a connection to Redis

In addition to using SQLite as the main database for our app, we also need to configure the connection to Redis so that we can use Laravel's queuing feature to broadcast our data to the Redis server, which will then be picked up by the Laravel Echo Server so that the event will be sent to the Vue 3 frontend. The environment variables for the Redis configuration are as follows:

```
BROADCAST_DRIVER=redis
QUEUE_CONNECTION=redis
QUEUE_DRIVER=sync
```

And we add the Redis configuration with the following:

```
REDIS_HOST=127.0.0.1
REDIS_PASSWORD=null
REDIS_PORT=6379
```

The first group of environment variables configures where the queue directs the data to. The `BROADCAST_DRIVER` setting is set to `redis` so that we direct our event to Redis. `QUEUE_CONNECTION` also has to be set to `redis` for the same reason. `QUEUE_DRIVER` is set to `sync` so that the events will be sent to the queue immediately after they are broadcast.

Running the migration files

Now that we have created our migrations and configured which database to use, we run `php artisan migrate` to run the migrations. Running the migrations will add the tables to our SQLite database. After adding the tables, we can add the seed data so we will not have to recreate the data ourselves when we want to reset our database or when we have an empty database. To create the seed data, we add some code to the `database/seeders/DatabaseSeeder.php` file. In the file, we write the following code to add the files for our database:

```php
<?php

namespace Database\Seeders;

use Illuminate\Database\Seeder;
use Illuminate\Support\Facades\DB;
use Illuminate\Support\Facades\Hash;
use Illuminate\Support\Str;
use App\Models\User;
use App\Models\Chat;

class DatabaseSeeder extends Seeder
{
    public function run()
    {
        $this->addUsers();
        $this->addChats();
        $this->addMessages();
    }

    private function addUsers()
    {
        for ($x = 0; $x <= 1; $x++) {
            DB::table('users')->insert([
                'name' => 'user'.$x,
                'email' => 'user'.$x.'@gmail.com',
                'password' => Hash::make('password'),
            ]);
```

```
                }
        }



        ...
}
```

We have the addUsers() method to add a few users into the users table. We create a loop that calls DB::table('users')->insert to insert some entries into the users table. The -> symbol is the same as the period in JavaScript. It lets us access object properties or methods.

In the insert() method, we pass in an associative array or dictionary with the keys and values that we want to insert:

```
    ...
    private function addChats()
    {
        for ($x = 0; $x <= 1; $x++) {
            DB::table('chats')->insert([
                'name' => 'chat '.$x,
            ]);
        }
    }
    ...
```

The addChats() method lets us add the chat room entries. We only have to insert the name. In the addMessages() method, we insert the entries for the messages table. We get the user entry's id value that we want to set as the value from an existing entry in the users table. Likewise, we do the same for chat_id by getting an entry from the chats table and use the id value for that entry and set that as the value of chat_id:

```
    ...
    private function addMessages()
    {
        for ($x = 0; $x <= 1; $x++) {
            DB::table('messages')->insert([
                'message' => 'hello',
                'user_id' => User::all()->get(0)->id,
```

```
                    'chat_id' => Chat::all()->get($x)->id
        ]);


        DB::table('messages')->insert([
            'message' => 'how are you',
            'user_id' => User::all()->get(1)->id,
            'chat_id' => Chat::all()->get($x)->id
        ]);
    }
  }
...
```

Once we have written the seeder, we may want to regenerate Composer's autoloader to update the autoloader with the dependencies that we have. We can do that by running `composer dump-autoload`. This is handy in case references to any dependencies are outdated and we want to refresh the references so that they won't be outdated. Then we run `php artisan db:seed` to run the seeder to populate all the data into the tables.

To reset the data to a pristine state, we can run the migration and the seeder at the same time by running `php artisan migrate:refresh –seed`. We can also just empty the database and rerun all the migrations by running `php artisan migrate:refresh`.

Creating our application logic

Now that we have got the database structure and the seed data down, we can move on to creating our app logic. We create some controllers so that we can receive requests in from the frontend. Laravel controllers should be in the `app/Http/Controllers` folder. We create one for receiving requests or manipulating the `chats` table and another for receiving requests to manipulate the `messages` table. Laravel comes with a command for creating controllers. First, we create the `ChatController.php` file by running the following code:

```
php artisan make:controller Chat
```

Then we should get the `app/Http/Controllers/ChatController.php` file added to our project. The full code is at `https://github.com/PacktPublishing/-Vue.js-3-By-Example/blob/master/Chapter08/backend/app/Http/Controllers/ChatController.php`.

A Laravel controller has a class that inherits from the `Controller` class. Inside the class, we have the methods that will be mapped to URLs later so that we can run these methods and do what we want. Each method takes a request object with the request data, including the headers, URL parameters, and body.

The `get()` method finds a single `Chat` entry. `Chat` is the model class for the `chats` table. In Laravel, the convention is that the class name corresponds to the table name by transforming the table name by removing the *s* at the end and then converting the first letter to uppercase. Therefore, the `Chat` model class is used to manipulate the entries in the `chats` table. Laravel does the mapping automatically so we don't have to do anything ourselves. We just have to remember this convention, so we won't get confused by it. The `find()` method is a `static` method that we use to get a single entry by its ID.

In all controller functions, we can just return a string, an associative array, a response object, or the results returned from a `query()` method to return that as the response. Therefore, the return value of the `Chat::find` method will be returned as the response when we make a request and the `get` method is called.

The `getAll()` method is used to get all the entries from the `chats` table. The `all()` method is a static method that returns all entries.

The `create()` method is used to create an entry from the request data. We call the `Validate::make` static method to create a validator for the request data. The first argument is `$request->all()`, which is a method that returns all the items in a request object. The second argument is an associative array with the key of the request body to validate. The value of it is a string with the validation rules. The required rule makes sure the name is filled. The string rule checks that value set as the value of the name key is a string. The `max:255` rule is the maximum number of characters we can have in the name value:

```
...
    public function create(Request $request)
    {
        $validator = Validator::make($request->all(), [
            'name' => 'required|string|max:255',
        ]);

        if($validator->fails()){
            return response()->json($validator->errors()-
```

```
            >toJson(), 400);
    }


    $chat = Chat::create([
        'name' => $request->get('name'),
    ]);


    return response()->json($chat, 201);
  }
...
```

We check whether validation fails with the $validator->fails() method.
$validator is the object that is returned by the Validator::make method. In the
if block, we call response()->json() to return the error to the user with the 400
status code.

Otherwise, we call Chat::create to create the chats table entry. We get the value of
the name field from the request body with the $request->get method with the key
that we want to get. Then we set that as the value of the 'name' key in the associative
array that we pass into the create method.

We do something similar with the update() method, except that we call Chat::find
to find the item by its id value. Then we assign the value of the name field from the
request body to the name property of the returned chat object. Then we call $chat-
>save() to save the latest value. Then we return the response by calling response()-
>json($chat) to return the latest chat entry converted to JSON:

```
...
    public function update(Request $request)
    {
        $validator = Validator::make($request->all(), [
            'name' => 'required|string|max:255',
        ]);


        if($validator->fails()){
            return response()->json($validator->errors()-
                >toJson(), 400);
        }
```

```
        $chat = Chat::find($request->id);
        $chat->name =   $request->get('name');
        $chat->save();

        return response()->json($chat);
    }
...
```

The `delete()` method is called when we make a *DELETE* request to the API to delete a chat room entry. We call `Chat::find` again to find the entry from the `chats` table with the given ID. Then we call `$chat->delete()` to delete the returned entry. Then we return an empty response:

```
...
    public function delete(Request $request)
    {
        $chat = Chat::find($request->id);
        $chat->delete();
        return response(null, 200);
    }
...
```

We have similar logic for `MessageController.php` to let us save chat messages. We have the `UserController.php` file with code to let us save user data when we register for a user account.

> **Important:**
>
> These files can be found at `https://github.com/PacktPublishing/-Vue.js-3-By-Example/blob/master/Chapter08/backend/app/Http/Controllers/MessageController.php` and `https://github.com/PacktPublishing/-Vue.js-3-By-Example/blob/master/Chapter08/backend/app/Http/Controllers/UserController.php` respectively.

Exposing controller methods for endpoints

Next, we have to map our controller methods to the URLs that we will make our requests to call. We do that by adding some code to the `routes/api.php` file. To do that, we replace what is in the file with the following code:

```php
<?php

use Illuminate\Http\Request;
use Illuminate\Support\Facades\Route;
use App\Http\Controllers\AuthController;
use App\Http\Controllers\UserController;
use App\Http\Controllers\ChatController;
use App\Http\Controllers\MessageController;
Route::post('register', [UserController::class, 'register']);
Route::group([
    'middleware' => 'api',
    'prefix' => 'auth'
], function () {
    Route::post('login', [AuthController::class, 'login']);
    Route::post('logout', [AuthController::class,
      'logout']);
    Route::post('refresh', [AuthController::class,
      'refresh']);
    Route::post('me', [AuthController::class, 'me']);
    });
...
    Route::get('{chat_id}', [MessageController::class,
      'getAll']);
    Route::post('create', [MessageController::class,
      'create']);
});
```

We expose the controller methods as POST and GET endpoints to the client by calling the `Route::post` and `Route::get` methods respectively.

The `jwt.verify` middleware is what we use to check the JSON web token before we run the `controller` method for the route. This way, the `controller()` method will only be run when the token is valid.

Then we have to create `AuthController` to let us do the JSON web token authentication.

First, we run the following command:

```
php artisan make:controller AuthController
```

Then in the `app/Http/Controllers/AuthController.php` file, we add methods for endpoints to let us get the current user's data, log in, and log out. The code for this file is at `https://github.com/PacktPublishing/-Vue.js-3-By-Example/ blob/master/Chapter08/backend/app/Http/Controllers/ AuthController.php`.

If you don't have the `app/Http/Middleware/JwtMiddleware.php` file, the full code for this file is at `https://github.com/PacktPublishing/-Vue.js-3- By-Example/blob/master/Chapter08/backend/app/Http/Middleware/ JwtMiddleware.php`.

It lets us enable authentication with a JSON web token between the Vue client and this app.

Setting up JWT authentication

Now we have to set up the JSON web token authentication with our Laravel app so that we can host our frontend separately from our backend. To do that, we use the `tymon/ jwt-auth` library. To install it, we run the following command:

```
composer require tymon/jwt-auth
```

Next, we run the following command to publish the package configuration files:

```
php artisan vendor:publish --provider="Tymon\JWTAuth\Providers\ LaravelServiceProvider"
```

The preceding command will add all the required configuration for us. We should now have `config/jwt.php` added to our app. Then we generate the secret key to sign the JSON web token by running the following command:

```
php artisan jwt:secret
```

The `secret` key will be added to the `.env` file with the key `JWT_SECRET`.

Configuring our authentication

Next, we have to configure our authentication so that we can verify our JSON web token before we can successfully make a request to routes that requires authentication. In `config/auth.php`, we have the following code:

```php
<?php

return [
    'defaults' => [
        'guard' => 'api',
        'passwords' => 'users',
    ],
    'guards' => [
        'web' => [
            'driver' => 'session',
            'provider' => 'users',
        ],

        'api' => [
            'driver' => 'jwt',
            'provider' => 'users',
        ],
    ],
    'providers' => [
        'users' => [
            'driver' => 'eloquent',
            'model' => App\Models\User::class,
        ],
    ],
    'passwords' => [
        'users' => [
            'provider' => 'users',
            'table' => 'password_resets',
            'expire' => 60,
            'throttle' => 60,
        ],
```

```
    ],
    'password_timeout' => 10800,
];
```

In the `guards` section, we have an `api` key that has a value being an associative array with the driver key set to `'jwt'` and the provider set to `'users'` to let us authenticate users with the JSON web token issued by the `jwt-auth` library.

Next, we add the code to enable CORS so that our Vue.js 3 app can communicate with it.

Enabling cross-domain communication

To enable us to do cross-domain communication between the frontend and backend, we install the `fruitcake/laravel-cors` package. To do this, we run the following command:

```
composer require fruitcake/laravel-cors
```

Then, in `app/Http/Kernel.php`, we should have the following:

```php
<?php

namespace App\Http;

use Illuminate\Foundation\Http\Kernel as HttpKernel;

class Kernel extends HttpKernel
{
    protected $middleware = [
        \App\Http\Middleware\TrustProxies::class,
        \Fruitcake\Cors\HandleCors::class,
        \App\Http\Middleware\
            PreventRequestsDuringMaintenance::class,
        \Illuminate\Foundation\Http\Middleware\
            ValidatePostSize::class,
        \App\Http\Middleware\TrimStrings::class,
        \Illuminate\Foundation\Http\Middleware\
            ConvertEmptyStringsToNull::class,
```

```
            \Fruitcake\Cors\HandleCors::class,
    ];


    ...
            'password.confirm' =>   \Illuminate\Auth\Middleware\
                RequirePassword::class,
            'signed' => \Illuminate\Routing\Middleware\
                ValidateSignature::class,
            'throttle' => \Illuminate\Routing\Middleware\
                ThrottleRequests::class,
            'verified' => \Illuminate\Auth\Middleware\
                EnsureEmailIsVerified::class,
            'jwt.verify' => \App\Http\Middleware\
                JwtMiddleware::class,
        ];
    }
```

We register the `HandleCors` middleware that comes with the `laravel-cors` package, and register the `jwt.verify` middleware in the `$routeMiddleware` associative array by writing the following code in the `$routesMiddleware` associative array:

```
'jwt.verify' => \App\Http\Middleware\JwtMiddleware::class,
```

This way, we can use the `jwt.verify` middleware to verify the token.

The full code is at `https://github.com/PacktPublishing/-Vue.js-3-By-Example/blob/master/Chapter08/backend/app/Http/Kernel.php`.

Also, we install the `predis` package so that we can talk to our Redis database more easily. To install the `predis` package, we run the following command:

```
composer require predis/predis
```

Then, in `config/database.php`, we write the following code:

```php
<?php

use Illuminate\Support\Str;

return [
```

```
...     'redis' => [

    'client' => env('REDIS_CLIENT', 'predis'),

    'options' => [
        'cluster' => env('REDIS_CLUSTER', 'redis'),
        'prefix' => env('REDIS_PREFIX', Str::slug(
        env('APP_NAME', 'laravel'), '_').'_database_'),
    ],

    ...

    'cache' => [
        'url' => env('REDIS_URL'),
        'host' => env('REDIS_HOST', '127.0.0.1'),
        'password' => env('REDIS_PASSWORD', null),
        'port' => env('REDIS_PORT', '6379'),
        'database' => env('REDIS_CACHE_DB', '1'),
    ],

    ],

];
```

We configure our Redis database connection within the associative array we set for 'redis' so that we can connect to Redis.

The full file is at https://github.com/PacktPublishing/-Vue.js-3-By-Example/blob/master/Chapter08/backend/config/database.php.

Now that we have created the API for storing users' data and their chat messages, we'll move on to add real-time communication capabilities to the Laravel app so that users can save and get chat messages in real time.

Adding real-time communication

Now that we are done with adding the routes, authentication, and database configuration and manipulation code, we are ready to add the code for letting us communicate between the frontend and backend in real time. First, we need to create an event class in our Laravel backend so that we can call the event function to broadcast the event as we did in MessageController.

To do that, we run the php artisan make:event MessageSent command to create the MessageSent event class. The class should now be in the backend/app/ Events/MessageSent.php file. Once the file is created, we replace what is inside the file with the following code:

```php
<?php

namespace App\Events;

...
class MessageSent implements ShouldBroadcast
{
    use InteractsWithSockets, SerializesModels;

    public $user;

    public $message;

    public function __construct(User $user, Message
      $message)
    {
        $this->user = $user;
        $this->message = $message;
    }

    public function broadcastOn()
    {
        return new Channel('chat');
    }
}
```

```php
    public function broadcastAs()
    {
        return 'MessageSent';
    }
}
```

The __constructor() method is the constructor; we get the $user and $message objects and then assign them to the class variables with the same name. The broadcastOn() method returns the Channel object, which creates a channel that we can listen to in the frontend. The broadCastAs() method returns the event name that we listen to in the chat channel. We will use this in the frontend to listen to the broadcast events. An event class should implement the ShouldBroadcast interface so that events can be broadcast from it.

The full code of MessageSent.php is at https://github.com/ PacktPublishing/-Vue.js-3-By-Example/blob/master/Chapter08/ backend/app/Events/MessageSent.php.

In the backend/routes/channels.php file, we should have the following code so that all users can listen to the chat channel:

```php
<?php

use Illuminate\Support\Facades\Broadcast;
Broadcast::channel('chat', function () {
    return true;
});
```

The first argument is the name of the channel we are subscribing to. The callback is a function that returns true if the user can listen to the event and false otherwise. An optional parameter is the user object so that we can check whether the user can listen to a given event.

The full code of this file is at https://github.com/PacktPublishing/-Vue. js-3-By-Example/blob/master/Chapter08/backend/routes/channels. php.

The communication flow is as shown in the following diagram:

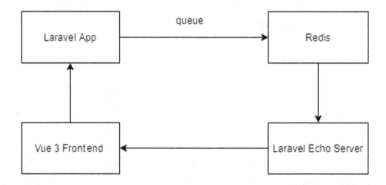

Figure 8.1 – Diagram of the chat app's architecture

The Vue 3 frontend makes an HTTP request to the Laravel app with the message we want to send. The Laravel app saves the message to the messages table with the ID of the chat session and the user. It also broadcasts an event via the queue to the Redis server. Then the Laravel Echo Server watches Redis to see whether there is anything saved to the Redis database. If there is anything new saved, then the Laravel Echo Server pushes that to the Vue 3 frontend. The Vue 3 frontend picks that up by listening to the Laravel Echo Server for the events with the Laravel Echo client and the Socket.IO client.

Communication with Socket.IO

To make our backend app communicate with the frontend via Socket.IO, we need the Laravel Echo Server. To do this, we first need to install the Laravel Echo Server npm package globally. We install it by running npm install -g laravel-echo-server. Then we will run this package to create the configuration file for setting up the communication.

To do this, we create a new folder and then run laravel-echo-server init to run the command-line wizard to create the Laravel Echo Server configuration file in the folder. At this point, we can just answer all the questions that are asked with the default settings. This is because we are going to edit the configuration file that it creates once this wizard is done.

Once the wizard is done, we should see the laraval-echo-server.json file in the folder. Now we open it and replace whatever is in there with the following code:

```
{
  "authHost": "http://localhost:8000",
  "authEndpoint": "/broadcasting/auth",
```

```json
  "clients": [
    {
      "appId": "APP_ID",
      "key": "c84077a4dabd8ab2a60e51b051c9d0ea"
    }
    ...
  },
  "sqlite": {
    "databasePath": "/database/laravel-echo-server.sqlite"
  },
  "publishPresence": true
},
"devMode": true,
"host": "127.0.0.1",
...
"http": true,
  "redis": true
},
"apiOriginAllow": {
  "allowCors": true,
  "allowOrigin": "*",
  "allowMethods": "GET, POST",
  "allowHeaders": "Origin, Content-Type, X-Auth-Token,
    X-Requested-With, Accept, Authorization, X-CSRF-TOKEN,
    X-Socket-Id"
  }
}
```

In the preceding code, we have the JSON code for the configuration so that the Laravel Echo Server can listen to the items that are saved in Redis and then send whatever is in the Redis database to the frontend via the Socket.IO client. The devMode property is set to true so that we can see all the events that are sent. The host has the host IP address of the Laravel Echo Server. The port property is set to 6001 so this server will listen to port 6001. Another important part of this file is the apiOriginAllow property. It is set to an object with allowCors set to true so that we can do cross-domain communication with our frontend.

The `allowOrigin` property lets us set the domains that are allowed to listen to the emitted events. The `allowMethods` property has the HTTP methods that are allowed to be received from the frontend. The `allowHeaders` property has the list of HTTP request headers that are allowed to be sent from the frontend to the Laravel Echo Server.

`authHost` has the base URL of the Laravel app so that it can listen to the events broadcast by the Laravel app. `authEndpoint` has the authentication endpoint that is used to check whether the user is authenticated for listening to events that require authentication.

Another important part of this configuration file is the database configuration properties. The database property is set to `"redis"` so that it will listen to the Redis server for saved items. The `databaseConfig` property has the settings to let us connect to the Redis server. The `"redis"` property is set to an object with the `"port"` property set to the `port` that the Redis server listens to. The default port for Redis is `6379`. The `host` property is the location of the Redis server. The `publishPresence` property is set to `true` so that Redis publishes items that are saved in its database.

The full configuration is at `https://github.com/PacktPublishing/-Vue.js-3-By-Example/blob/master/Chapter08/laravel-echo-server/laravel-echo-server.json`.

The latest version of Redis is only available for Linux or macOS. To install Redis on Ubuntu Linux, run the following command to install the Redis server:

```
sudo apt update
sudo apt install redis-server
```

If you are running Windows 10, you can use the Window Subsystem for Linux to install a copy of Ubuntu Linux so you can run the latest version of Redis. To install Ubuntu on Windows 10, do the following:

1. Type in `Turn Windows Features on and Off` in the **Start** menu.

    Then we scroll to the bottom and click on **Window Subsystem for Linux** to install it. It will ask you to restart, and you should do that before continuing.

2. Once your computer has restarted, go to the **Windows Store** and search for **Ubuntu**, then you can click it and click **Get**.

3. After it is installed, then you can type in `Ubuntu` in the **Start** menu and start it. Now just follow the instructions to finish the installation.

Then you can run the preceding two commands to install Redis.

Once Redis is installed, we run the following command to run the Redis server:

```
redis-server
```

Now the backend part of our project is done. Now we run `php artisan serve` and `php artisan queue:listen` to run the Laravel app and the queue worker. We also have to run the Laravel Echo Server by running `laravel-echo-server start` to start the Laravel Echo Server.

If you run into any issues, then you may have to clear the cache to make sure that the latest code is actually running. To do that, you can run the following commands to clear all the caches:

```
php artisan config:cache
php artisan config:clear
php artisan route:cache
php artisan route:clear
```

If the cache is cleared and the code is still not working, then you can go back to check the code.

Now that we have added real-time communication to our Laravel app, we are ready to move on to creating the frontend to let users register an account, log in, and start chatting in a chat room.

Creating the frontend to let users chat

Now that we have the backend code all finished and running, we can work on the frontend. The frontend is not very different from what we had in the earlier chapters. We use the Vue CLI to create our project in the `frontend` folder of the `vue-example-ch8-chat-app` folder and then we can start writing our code.

In the `vue-example-ch8-chat-app/frontend` folder, we run `vue create`, then we choose **select version**, then we select the **Vue 3** option with **Vue Router** option enabled. Once the Vue CLI wizard finishes running, we can start building our frontend.

Installing the Vue dependencies

In addition to the Vue dependencies, we also need to install the Axios HTTP client, the Socket.IO client, and the Laravel Echo client package to make HTTP requests and listen to events emitted from the server side via the Laravel Echo Server respectively. To install those, we run the following command:

```
npm install axios socket.io-client laravel-echo
```

First, in the src folder, we create the constants.js file and add the following code:

```
export const APIURL = 'http://localhost:8000';
```

We add the APIURL constant that we will use when we make requests to our API endpoints. In src/main.js, we replace the code we have with the following:

```
...

axios.interceptors.request.use((config) => {
  if (config.url.includes('login') ||
  config.url.includes('register')) {
    return config;
  }
  return {
    ...config, headers: {
      Authorization: `Bearer ${localStorage.getItem('token')}`,
    }
  }
}, (error) => {
  return Promise.reject(error);
});

axios.interceptors.response.use((response) => {
  const { data: { status } } = response;
  if (status === 'Token is Expired') {
    router.push('/login');
  }
  return response;
}, (error) => {
  return Promise.reject(error);
```

```
});
```

```
createApp(App).use(router).mount('#app')
```

We have two things that are new in this file. We have the Axios request and response interceptors so that we can apply the same settings to all requests without repeating the same code when we make each request. The `axios.interceptors.request.use()` method takes a callback that lets us return a new `config` object according to our needs.

If the request URL doesn't include `login` or `register`, then we need to add the token to the `Authorization` header. This is what we did in the callback we passed into the `use()` method. We add the token to the request configuration for the endpoints that need them. The second callback is an error handler, and we just return a rejected promise so we can handle them when we make the request.

Similarly, we have the `axios.interceptor.response.use()` method to check for each response with the callback in the first argument. We check whether the response body has the `status` property set to the `"Token is expired"` string so we can redirect to the login page when we get this message and return the response. Otherwise, we just return the response as is. The error handler in the second argument is the same with the request interceptor.

Creating our components

Next, we create our components. We start with the form to let us set or edit the chat room name. To do that, we go into the `src/components` folder and create the `ChatroomForm.vue` file. Then, in the file, we write the following code:

```
<template>
  <div>
    <h1>{{ edit ? "Edit" : "Add" }} Chatroom</h1>
    <form @submit.prevent="submit">
      <div class="form-field">
        <label for="name">Name</label>
        <br />
        <input v-model="form.name" type="text" name="name"
          />
      </div>
      <div>
        <input type="submit" />
      </div>
```

```
    </form>
  </div>
</template>
. . .
```

This component takes the `edit` prop, which has the type boolean and the `id` prop that has the type string. It has one reactive property, which is the `form` property. It is used to bind the input value to the reactive property. We have the `submit()` method that checks the name to see whether it is filled. If it is, then we go ahead and submit it. If the `edit` prop is true, then we make a PUT request to update an existing entry in the `chats` table with the given ID. Otherwise, we create a new entry in the same table with the given name value. Once that is done, we redirect to the home page, which has the list of chat rooms.

In the `created` hook, we check whether the `edit` reactive property is true. If it is, then we get the entry in the `chats` table with the given ID and set it as the value of the `form` reactive property so that we can see the value of the `form.name` property in the input box:

```
<script>
import axios from "axios";
import { APIURL } from "../constants";
export default {
  name: "ChatroomForm",
  . . .
  async created() {
    if (this.edit) {
      const { data } = await
      axios.get(`${APIURL}/api/chat/${this.id}`);
      this.form = data;
    }
  },
};
</script>
```

Next, in the `src/components` folder, we create `NavBar.vue` to create a component to render a navigation bar. Inside the file, we write the following code:

```
<template>
  <div>
    <ul>
      <li>
```

```
            <router-link to="/">Chatrooms</router-link>
        </li>
        <li><a href="#" @click="logOut">Logout</a></li>
    </ul>
  </div>
</template>

<script>
import axios from "axios";
import { APIURL } from "../constants";

export default {
  name: "NavBar",
  methods: {
    async logOut() {
      await axios.post(`${APIURL}/api/auth/logout`);
      localStorage.clear();
      this.$router.push("/login");
    },
  },
};
</script>
...
```

We have a `router-link` component that goes to the **Chatrooms** page. This is done by setting the `to` prop to the `/` route. We also have a link to call the `logout()` method when we click it. The `logout()` method makes a POST request to the `/api/auth/logOut` endpoint to invalidate the JSON web token. Then we call the `localStorage.clear()` method to clear local storage. Then we call `this.$router.push` to redirect to the login page.

In the styles section, we have some styles for the `ul` and `li` elements so the `li` display horizontally with some margins between them. We also have the `list-style-type` property set to `none` so that we remove the bullets from the list:

```
<style scoped>
ul {
  list-style-type: none;
```

```
}

ul li {
  display: inline;
  margin-right: 10px;
}
</style>
```

In the `src/views` folder, we have the components for the pages that we map to a URL with Vue Router so that we can access these components from the browser. First, we create the `AddChatroomForm.vue` component file in the `src/views` folder and add the following code:

```
<template>
  <div>
    <NavBar></NavBar>
    <ChatroomForm></ChatroomForm>
  </div>
</template>

<script>
import ChatroomForm from "@/components/ChatroomForm";
import NavBar from "@/components/NavBar";

export default {
  components: {
    ChatroomForm,
    NavBar
  },
};
</script>
```

We just register the `NavBar` and the `ChatroomForm` component in the `components` property and then add them to the template.

Next, we create the `ChatRoom.vue` component to display our chat messages and add the code to listen to the `laravel_database_chat` channel's `MessageSent` event that we emit from the Laravel app via the Redis database and the Laravel Echo Server in this file. In this file, we write the following code:

```
...
<script>
import axios from "axios";
import { APIURL } from "../constants";
import NavBar from "@/components/NavBar";

export default {
  name: "Chatroom",
  components: {
    NavBar,
  },
  beforeMount() {
    this.getChatMessages();
    this.addChatListener();
  },
  data() {
    return {
      chatMessages: [],
      message: "",
    };
  },
  ...};
</script>
```

Then we add the methods to the same file to get and send the chat messages by writing the following code:

```
...
<script>
...
export default {
  ...
  methods: {
    async getChatMessages() {
      const { id } = this.$route.params;
      const { data } = await
        axios.get(`${APIURL}/api/message/${id}`);
```

```
        this.chatMessages = data;
        this.$nextTick(() => {
          const container = this.$refs.container;
          container.scrollTop = container.scrollHeight;
        });
      },
      async sendChatMessage() {
        const { message } = this;
        if (!message) {
          return;
        }
        const { id: chat_id } = this.$route.params;
        ...
          () => {
            this.getChatMessages();
          }
        );
      },
    },
```

```
</script>
```

The getChatMessages method gets the chat messages for the chat room from the API and the sendChatMessage method sends a message to the chatroom by submitting the chat message via an HTTP request to the API. Then, the API endpoint would send a message to the queue through the Laravel Echo Server back to the Socket.IO chat client used in this app. We call addChatListener to listen to the laravel_database_ chat event from the server, which calls getChatMessages to get the latest messages.

The component template just uses the v-for directive to render each entry of the chatMessages reactive property and render them. The form element in the following code is used to let us enter a message and then submit it to Laravel by making an HTTP request. The endpoint saves the message to the messages table and also emits an event that we listen to, which is sent via the Redis database and the Laravel Echo Server. The frontend only knows about the Laravel Echo Server from the real-time communication point of view:

```
<template>
  <div>
    <NavBar></NavBar>
```

```
<h1>Chatroom</h1>
<div id="chat-messages" ref="container">
  <div class="row" v-for="m of chatMessages"
    :key="m.id">
    <div>
      <b>{{ m.user.name }} - {{ m.created_at }}</b>
    </div>
    <div>{{ m.message }}</div>
  </div>
</div>
<form @submit.prevent="sendChatMessage">
  <div class="form-field">
    <label for="message">Message</label>
    <br />
    <input v-model="message" type="text" name="message"
      />
  </div>
  <div>
    <input type="submit" />
  </div>
</form>
</div>
</template>
```

In the component object, we have the `beforeMount` hook to call the `getChatMessage` method to get chat messages. The `addChatListener()` method creates an event listener with the Socket.IO client to let us listen to events emitted from the Laravel Echo Server. In the `getChatMessage()` method, we call the `this.$nextTick()` method with a callback so that we always scroll to the bottom of the `div` tag that holds the messages once we get the messages. We have run that code in the `$nextTick` callback because we need to make sure the scrolling code runs after all the messages are rendered.

The `this.$nextTick()` method lets us wait for the component to be re-rendered after reactive properties are updated before running the code in the callback.

In the `addChatListener()` method, we subscribe to the `laravel_database_chat` channel, which is the same as the chat channel we defined in the Laravel app. We can make sure that we subscribe to the right channel by watching the output of the Laravel Echo Server. The `.MessageSent` event is the same one that we defined in the backend app. The dot before the event name is required so that it listens to the right event in the right namespace. In the callback that we pass in to listen, we call `this.getChatMessages()` to get the latest messages.

The container for the chat messages is set to the height 300px so that it won't be too tall when we have too many messages. It also lets us scroll to the bottom when we have enough messages to overflow the container:

```
<style scoped>
#chat-messages {
  height: 300px;
  overflow-y: scroll;
}

.row {
  display: flex;
  flex-wrap: wrap;
}

.row div:first-child {
  width: 30%;
}
</style>
```

Next, in the `src/views` folder, we create the `Chatrooms.vue` component file by writing the following code:

```
<template>
  <div>
    <NavBar></NavBar>
    <h1>Chatrooms</h1>
    <button @click="createChatRoom">Create Chatroom
      </button>
    <table id="table">
      <thead>
```

```
        <tr>
            <th>Name</th>
            <th>Go</th>
            <th>Edit</th>
            <th>Delete</th>
...
    beforeMount() {
        this.getChatRooms();
    },
};
</script>

<style scoped>
#table {
    width: 100%;
...
</style>
```

We render a table with the list of chat rooms that we can go to, edit the name of, or delete. The method just gets the chat room data and goes to the routes for editing the chatroom with the given ID, adding the chat room, redirecting to the chat room page with the given ID, and deleting the chat room respectively. When we delete a chat room, we get the latest entries again with the `getChatRooms()` method so that we can get the latest data.

We get the chat room list in the `beforeMount` hook so we see the table entries when the page loads. Next, in the same folder, we create the `EditChatroomForm.vue` file and add the following code:

```
<template>
    <div>
        <NavBar></NavBar>
        <ChatroomForm edit :id="$route.params.id">
        </ChatroomForm>
    </div>
</template>

<script>
import ChatroomForm from "@/components/ChatroomForm";
```

```
import NavBar from "@/components/NavBar";

export default {
  components: {
    ChatroomForm,
    NavBar,
  },
};
</script>
```

It has the same content as the `AddChatroomForm.vue` file but with the `edit` prop on `ChatroomForm` set to `true` and the `id` prop set to the `id` URL parameter from the URL that we get from Vue Router.

Creating the login page

Next, we create the login page by creating `src/views/Login.vue` and adding the following code:

```
<template>
  <div>
    <h1>Login</h1>
    <form @submit.prevent="login">
      <div class="form-field">
        <label for="email">Email</label>
        <br />
        <input v-model="form.email" type="email" name="email"
/>
      </div>
      <div class="form-field">
        <label for="password">Password</label>
        <br />
        <input v-model="form.password" type="password"
            name="password" />
      </div>
      <div>
        <input type="submit" value="Log in" />
        <button type="button" @click="goToRegister">Register</
```

```
              button>
        </div>
      </form>
    </div>
  </template>

  <script>
  import axios from "axios";
  import { APIURL } from "../constants";

  export default {
    name: "Login",
    data() {
      return {
        form: {
          email: "",
          password: "",
        },
      };
    },
    methods: {
      async login() {
        const { email, password } = this.form;
        if (!email || !password) {
          alert("Email and password are required");
          return;
        }

        try {
          const {
            data: { access_token },
          } = await axios.post(`${APIURL}/api/auth/login`, {
            email,
            password,
          });
          localStorage.setItem("token", access_token);
```

```
            this.$router.push("/");
        } catch (error) {
            alert("Invalid username or password");
        }
    },
    goToRegister() {
        this.$router.push("/register");
    },
    },
};
</script>
```

The template just has a login form that takes the email and password to let us log in. When we submit the form, the `login()` method is called. It first checks whether all the fields are filled in properly, then it makes an HTTP request to the `/api/auth/login` route with the credentials to see whether we can log in.

The form also has a `Register.vue` file to go to the **Register** page so that we can register for an account and join the chat rooms.

The last page that we have to create is a page for housing the registration form. To create it, we write the following code:

```
<template>
  <div>
    <h1>Register</h1>
    <form @submit.prevent="register">
      <div class="form-field">
        <label for="email">Name</label>
        <br />
        <input v-model="form.name" type="text" name="name"
          />
      </div>
      <div class="form-field">
        <label for="email">Email</label>
        <br />
        <input v-model="form.email" type="email"
          name="email" />
      </div>
```

```
        <div class="form-field">
          <label for="password">Password</label>
          <br />
            ...
            name,
            email,
            password,
            password_confirmation: confirmPassword,
          });
          this.$router.push("/login");
        } catch (error) {
          alert("Invalid username or password");
        }
      },
    },
  };
</script>
```

The form has the **Name**, **Email**, **Password**, and **Confirm Password** fields that are all required to register an account. When we submit the form, we call the `register()` method. We do the checks for the fields to see whether they're filled out properly. The `test()` method is called on the `email` regular expression to check for a valid email address. If it is valid, the `test()` method returns `true`. Otherwise, it returns `false`. We also check whether the password is the same as the `confirmPassword` variable. If everything is valid, then we make a POST request to register for a user account.

In `src/App.vue`, we replace what is there with the following code to add the `router-view` component so that we can see the route components from the `src/views` folder:

```
<template>
  <div>
    <router-view />
  </div>
</template>

<style scoped>
div {
  width: 70vw;
```

```
     margin: 0 auto;
   }
</style>

<style>
.form-field input {
   width: 100%;
}
</style>
```

Then, in the `src/router/index.js` file, we replace what is there with the following code to register all the routes and also create the `Socket.io` event to listen with the Laravel Echo library:

```
...
window.io = require('socket.io-client');

const beforeEnter = (to, from, next) => {
   const hasToken = Boolean(localStorage.getItem('token'));
   if (!hasToken) {
     return next({ path: 'login' });
   }
   next();
}
const routes = [
   {
...
     path: '/edit-chatroom/:id',
     name: 'edit-chatroom/:id',
     component: EditChatroomForm,
     beforeEnter
   },
]
...
export default router
```

The Laravel Echo client is used with the Socket.IO client so that we can listen to events broadcast from the Laravel Echo Server. The `broadcaster` property is set to `'socket. io'` so that we can listen to the events from the Laravel Echo Server. The `host` property is set to the URL of the Laravel Echo Server.

Also, we have the `beforeEnter` navigation guard that we have seen in previous chapters for when we need to restrict a route to be available only after authentication is successful. We just check whether the token exists. If it exists, we call `next` to proceed to the next route. Otherwise, we redirect to the login page.

Now we can run the frontend by running `npm run serve` as we do in all the other projects. Now we should see something like the following screenshots. The following screenshot has the **Chatroom** user interface:

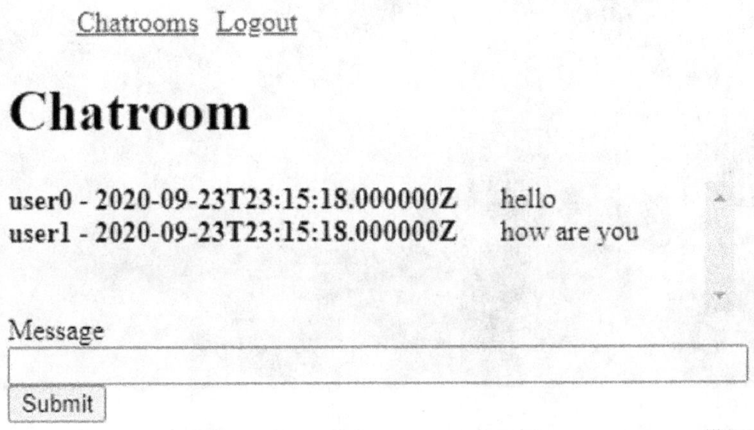

Chatrooms Logout

Chatroom

user0 - 2020-09-23T23:15:18.000000Z   hello
user1 - 2020-09-23T23:15:18.000000Z   how are you

Message

Submit

Figure 8.2 – Screenshot of the Chatroom

The following screenshot is the Laravel Echo Server working. We should see the name of the event that is broadcast and the channel that it is sent through:

```
laravel-echo-server                                                    —  □  ×

L A R A V E L   E C H O   S E R V E R

version 1.6.2

⬚ Starting server in DEV mode...

⬚  Running at 127.0.0.1 on port 6001
⬚  Channels are ready.
⬚  Listening for http events...
⬚  Listening for redis events...

Server ready!

[1:38:19 PM] - vT1nPvbkh_ySqLWuAAAM joined channel: laravel_database_chat
Channel: laravel_database_chat
Event: MessageSent
[1:39:54 PM] - vT1nPvbkh_ySqLWuAAAM left channel: laravel_database_chat (transport close)
[1:39:54 PM] - N7irdhIhjHvY0tPVAAAN joined channel: laravel_database_chat
[1:41:30 PM] - N7irdhIhjHvY0tPVAAAN left channel: laravel_database_chat (transport close)
[1:41:31 PM] - 7Oh3S9YgOoRPAy_LAAAO joined channel: laravel_database_chat
[1:42:25 PM] - 7Oh3S9YgOoRPAy_LAAAO left channel: laravel_database_chat (transport close)
[1:42:26 PM] - 0m1qwDRXIvEH04y_AAAP joined channel: laravel_database_chat
[1:43:03 PM] - 0m1qwDRXIvEH04y_AAAP left channel: laravel_database_chat (transport close)
[1:43:04 PM] - 8yA2AAAVrH56THkkAAAQ joined channel: laravel_database_chat
[1:43:47 PM] - 8yA2AAAVrH56THkkAAAQ left channel: laravel_database_chat (transport close)
[1:43:48 PM] - 13XIxpmP7S2LukhyAAAR joined channel: laravel_database_chat
[1:43:57 PM] - 13XIxpmP7S2LukhyAAAR left channel: laravel_database_chat (transport close)
[1:43:57 PM] - QMfSH43G6lbHjzwGAAAS joined channel: laravel_database_chat
[1:44:08 PM] - QMfSH43G6lbHjzwGAAAS left channel: laravel_database_chat (transport close)
```

Figure 8.3 – The output from Redis when chat events are sent to the frontend

The following screenshot is the log of the event of the queue:

```
C:\WINDOWS\system32\cmd.exe - php artisan queue:listen                  —  □  ×

c:\vue-example-ch8-chat-app\backend>php artisan queue:listen
[2020-09-23 23:14:35][g2SmHtIpeOAV2FrQarwZWfdzL9rl0vP7] Processing: App\Events\MessageSent
[2020-09-23 23:14:36][g2SmHtIpeOAV2FrQarwZWfdzL9rl0vP7] Processed:  App\Events\MessageSent
```

Figure 8.4 – Output of Laravel events

We started the queue earlier by running `php artisan queue:listen` in the backend folder, which is the folder with the Laravel project.

Now that we have the chat app's frontend working, we have created a simple chat system with Laravel and Vue.

Summary

In this chapter, we looked at how to build a chat app with Laravel and Vue. We built our backend with Laravel and we added controllers to receive requests. And we used the queue system built into Laravel to send data to the frontend. We also added JSON web token authentication into our Laravel app.

On the frontend, we used the Socket.IO client to listen to events sent from the Laravel Echo Server, which gets its data from Laravel via Redis.

Now that we have gone through Vue 3 projects with various levels of difficulty, we can adapt what we learned here to real-life situations. Real-life Vue apps will almost always make HTTP requests to a server. The Axios library makes this easy. Some apps also communicate in real time with the server like the chat app in this chapter.

The only difference is that in real-life apps, there would be checks to see whether the user is authenticated and authorized to send the data to the server.

`Packt.com`

Subscribe to our online digital library for full access to over 7,000 books and videos, as well as industry leading tools to help you plan your personal development and advance your career. For more information, please visit our website.

Why subscribe?

- Spend less time learning and more time coding with practical eBooks and Videos from over 4,000 industry professionals

- Improve your learning with Skill Plans built especially for you

- Get a free eBook or video every month

- Fully searchable for easy access to vital information

- Copy and paste, print, and bookmark content

Did you know that Packt offers eBook versions of every book published, with PDF and ePub files available? You can upgrade to the eBook version at `packt.com` and as a print book customer, you are entitled to a discount on the eBook copy. Get in touch with us at `customercare@packtpub.com` for more details.

At `www.packt.com`, you can also read a collection of free technical articles, sign up for a range of free newsletters, and receive exclusive discounts and offers on Packt books and eBooks.

Other Books You May Enjoy

If you enjoyed this book, you may be interested in these other books by Packt:

Building Vue.js Applications with GraphQL

Heitor Ramon Ribeiro

ISBN: 978-1-80056-507-4

- Set up your Vue.js projects with Vue CLI and explore the power of Vue components
- Discover steps to create functional components in Vue.js for faster rendering
- Become familiar with AWS Amplify and learn how to set up your environment
- Understand how to create your first GraphQL schema
- Use Quasar Framework to create simple and effective interfaces
- Discover effective techniques to create queries for interacting with data
- Explore Vuex for adding state management capabilities to your app
- Discover techniques to deploy your applications effectively to the web

Vue.js 3 Cookbook

Heitor Ramon Ribeiro

ISBN: 978-1-83882-622-2

- Design and develop large-scale web applications using Vue.js 3's latest features
- Create impressive UI layouts and pages using Vuetify, Buefy, and Ant Design
- Extend your Vue.js applications with dynamic form and custom rules validation
- Add state management, routing, and navigation to your web apps
- Extend Vue.js apps to the server-side with Nuxt.js
- Discover effective techniques to deploy your web applications with Netlify
- Develop web applications, mobile applications, and desktop applications with a single code base using the Quasar framework

Packt is searching for authors like you

If you're interested in becoming an author for Packt, please visit `authors.packtpub.com` and apply today. We have worked with thousands of developers and tech professionals, just like you, to help them share their insight with the global tech community. You can make a general application, apply for a specific hot topic that we are recruiting an author for, or submit your own idea.

Leave a review - let other readers know what you think

Please share your thoughts on this book with others by leaving a review on the site that you bought it from. If you purchased the book from Amazon, please leave us an honest review on this book's Amazon page. This is vital so that other potential readers can see and use your unbiased opinion to make purchasing decisions, we can understand what our customers think about our products, and our authors can see your feedback on the title that they have worked with Packt to create. It will only take a few minutes of your time, but is valuable to other potential customers, our authors, and Packt. Thank you!

Index

S

script tag
    used, for setting up Vue project 3, 4
shop item
    adding, with resolver function 214, 215
    getting 213, 214
shopping cart system project
    setting up 207
slider puzzle game
    components, creating for
        shuffling pictures 61-64
    score, calculating based on timing 69-71
    slides, rearranging 65-68
Socket.IO
    backend app, communicating
        to frontend 274-277
SQLite
    URL 222
SQLite database
    creating 222, 223
SQLite program
    download link 172, 259
Structured Query Language (SQL) 211

T

templates 2
template syntax 16, 17
testing 71
tests 53
TypeScript
    about 117
    benefits 118
    intersection types 118
    union types 118

U

Uniform Resource Locator (URL) 27, 225
unit testing
    with Jest 72-80
unit tests 53
Universal Module Definition (UMD) 4
user input
    handling 12, 13

V

vacation booking project
    admin frontend pages, creating 173-178
    creating 162, 163
vacation booking project, admin frontend
    creating 173
    Menubar component, adding 178-180
    shared code, adding to deal
        with requests 180-192
    top bar component, adding 178-180
vacation booking project, backend
    authentication middleware,
        adding 163, 164
    creating 163
    routes, adding to handle
        requests 164-172
vacation booking project, user frontend
    creating 193-200
    entry point code, adding 200-202
    router-view, adding 200-202
Vee-Validate 161
Vee-Validate 4 161
Vue 3
    core features 6